A HISTORY OF
JUNGLE
WARFARE

A HISTORY OF
JUNGLE
WARFARE

FROM THE EARLIEST DAYS TO THE BATTLEFIELDS OF VIETNAM

BRYAN PERRETT

FOREWORD BY
GENERAL SIR WALTER WALKER KCB CBE DSO

Pen & Sword
MILITARY

First published in Great Britain in 2015
and reprinted in paperback format in 2022 by
PEN AND SWORD MILITARY
An imprint of
Pen & Sword Books Ltd
Yorkshire – Philadelphia

ISBN 978 1 39902 016 9

A CIP catalogue record for this book is available from the British Library.

Typeset in the UK
Printed and bound in the UK by CPI Group (UK) Ltd.

Pen & Sword Books Limited incorporates the imprints of Atlas, Archaeology,
Aviation, Discovery, Family History, Fiction, History, Maritime, Military, Military
Classics, Politics, Select, Transport, True Crime, Air World, Frontline Publishing,
Leo Cooper, Remember When, Seaforth Publishing, The Praetorian Press,
Wharncliffe Local History, Wharncliffe Transport, Wharncliffe True Crime and
White Owl.

For a complete list of Pen & Sword titles please contact
PEN & SWORD BOOKS LIMITED
47 Church Street, Barnsley, South Yorkshire, S70 2AS, England
E-mail: enquiries@pen-and-sword.co.uk
Website: www.pen-and-sword.co.uk

Or

PEN AND SWORD BOOKS
1950 Lawrence Rd, Havertown, PA 19083, USA
E-mail: Uspen-and-sword@casematepublishers.com
Website: www.penandswordbooks.com

MIX
Paper | Supporting
responsible forestry
FSC
www.fsc.org
FSC® C013604

CONTENTS

ACKNOWLEDGEMENTS

The author wishes to express his sincere thanks for the generous assistance and advice so kindly provided by the following: Major Terence Gossage, Dr A. S. Binnie, Captain Isobe Takuo, Mr James Murray, Martin Windrow, and Simon Dunston.

FOREWORD

by

General Sir Walter Walker, KCB, CBE, DSO[3]

This book is a masterpiece for having condensed so vividly and accurately the many jungle warfare campaigns that have been fought from the earliest days of forest fighting to the present time when the helicopter has proved to be such a unique battle winner in the jungle.

The book has also succeeded in having extracted so accurately the main lessons from the type of fighting, the type of terrain and the type of climate that call for individual stamina and fortitude, stout legs, stout hearts, fertile brains, and the acceptance of battlefield conditions almost unimaginable in their demands on human endurance. The soldiers must be able to live in the jungle as close to the animal as it is humanly possible.

The tactical techniques and battle skills required would do credit to the tricks of the trade of a cat-burglar, gangster, gunman and poacher, for they require superb physical fitness, versatility, keen eyesight, silent movement, an eye for country, track discipline, concealment, surprise, marksmanship, guile, cunning and above all self-discipline.

The West's fundamental weakness has been that most of our military training and thinking has become focused on nuclear and conventional tactics for a European theatre against a first-class enemy. So when jungle warfare campaigns have broken out we have forgotten most of our jungle warfare techniques and expertise, learned the hard way and often at such high cost. We should learn our lessons from what went wrong at the beginning of a campaign and not from what went right at the end of it.

Today there are many flash-points and choke-points outside the Nato area where, if a conflagration were to break out, jungle warfare would be the order of the day. At present Terrorism is Public Enemy No 1. It is merely a new form of warfare but it demands many of the skills and principles evolved by those who have fought successful jungle campaigns.

One of the best books to come out of the Second World War, and regarded today as essential reading by Staff Colleges and Defence Institutions, is *Defeat Into Victory*, by the late Field Marshal Lord Slim of Burma. I will conclude this Foreword with two of his prophetic statements. He wrote:

'I believe that jungle fighting is today, strange as it may seem, the best training for nuclear war.'

He goes on to explain this by saying that formations will be compelled to disperse and that dispersed fighting will require skilled and determined junior leaders, and self-reliant, physically hard, well-disciplined troops. He ends with these words:

'In nuclear war, after the first shock of mutual devastation has been survived, victory will go, as it does in jungle fighting, to the tougher, more resourceful infantry soldier. . . The easier and more gadget-filled our daily life becomes, the harder will it be to produce him.'

INTRODUCTION
The Jungle Backdrop

The word jungle has its origins in Hindi and is derived from the Sanskrit *jangala*, meaning desert, although perhaps wilderness would be a more appropriate definition. In common parlance, however, it has come to mean tropical forest and specifically rain forests which receive in excess of 70 inches precipitation per annum. In primary jungle little light penetrates the thick overhead canopy of leaves and, because of this, the floor of the forest is both wet and soft and there is only sparse undergrowth. Where light does penetrate, for example around the edges of the forest or in areas which have been cleared and allowed to revert, secondary jungle consisting of dense, tangled vegetation flourishes. In equatorial rain forests such as those found in Africa the average day and night temperatures are respectively 86 degrees and 68 degrees, accompanied by high humidity. The sub-tropical rain forests found in the West Indies, in Vietnam and along the coast of Burma have a lower rainfall and since more light penetrates the canopy the undergrowth is thicker. In the monsoon rain forests covering much of Burma and South-East Asia there is a prolonged dry season which can produce drought conditions. Where the jungle extends to the coast mangrove trees entangle their roots, trapping mud and creating foetid swamps. Together, the topography and the jungle can form an impassable obstacle and in such circumstances rivers

provide the only possible highway. In suitable areas man can clear the jungle and harness the forces which created it to grow crops such as rice or rubber.

The jungle is never silent. Day and night animals and birds prey upon lesser breeds and each other for survival, shattering the green gloom with the chatter and shrieks of their life cycle. For man, however, the worst enemies are utterly soundless: together, such scourges as malaria, dysentery, cholera, beri beri, dengue fever, typhus and yellow fever have killed many times the number of those who have died in battle. Only in comparatively recent times was it discovered that there existed a connection between mild infection and subsequent immunity, and that the jungle often contained its own remedies. In the jungle, even the smallest insect bite can develop into a horribly infected sore, while leeches can penetrate the tiniest gap in boots and clothing to attach themselves to every part of the body, sucking blood until forced to relinquish their hold by the application of a cigarette end.

At the personal level, the techniques of jungle warfare are as old as the Stone Age. The club may have been replaced by the sophisticated assault rifle, but survival still depends upon acute observation, particularly of the unnatural, and extra reliance on the senses of smell and hearing. To those unfamiliar with the jungle, it is an alien environment pregnant with menace, apparently allied with an enemy skilled in the art of setting vicious traps which can kill or maim, offering endless opportunities for ambush. Acclimatization, familiarization and training all teach that the jungle, like the desert, is neutral, but the maintenance of morale hinges upon such factors as good junior leadership, regular supply and efficient casualty evacuation.

Some aspects of jungle warfare were apparent as early as AD 9, when Varus's legions were destroyed in the heavily wooded Teutoburger Wald. Others became evident during the forest fighting which formed so important a part of the eighteenth century campaigns in North America. More still were learned as the Western powers extended their influence into the tropics and found themselves conducting punitive expeditions or counter-insurgency operations. Yet jungle warfare as we understand it today, both high intensity between regular forces at the tactical, operative and strategic levels, and low intensity confrontation and counter-insurgency, has a comparatively short history dating from the Japanese invasion of Western territories in 1941-42. Since then, it is probably fair to say that more battles affecting the course of world history have been fought in the jungle than in any other environment.

Once it was thought that the jungle was the sole preserve of infantry supported by a few pack-artillery weapons. During and since the Second World War, however, every branch of service has learned to operate effectively there. The idea that tanks could exert a decisive influence, formerly regarded as idiotic, has been proved time and again. Improved radio communications have enabled the artillery to develop new tactics for attack and defence, and close support by ground-attack aircraft. Air-supply drops to forward troops have permitted them to maintain the momentum of an advance or operate in isolation, while earth-moving equipment has quickly converted forest trails into tracks passable by motor transport. Preventive medicine has reduced the risk from disease and pre-packaged rations have prolonged the soldier's ability to remain operational without re-supply. Since the Second World War the advent of the helicopter has provided the new dimension of air mobility and simplified casualty evacuation. Science has contributed a wide variety of sensor devices which can detect the presence of an enemy who could formerly rely upon remaining unseen.

Even so, despite the fact that the immense technical progress achieved in recent years has eased the soldier's burden considerably, few would wish to argue that jungle war remains one of the most gruelling of all forms of warfare and one which makes the severest demands on its participants.

CHAPTER 1

'DON'T NEVER TAKE A CHANCE YOU DON'T HAVE TO'

During the years immediately preceding the birth of Christ, Caesar Augustus decided to extend the northern frontier of the Roman Empire to the River Elbe. This was achieved against bitter resistance from the German tribes but by AD 5 the Emperor's adopted son, Tiberius, had quelled all opposition and imposed the Roman will on the now quiescent but sullenly resentful province.

The following year he was required to put down a serious revolt in Pannonia and was replaced by one Publius Quintilius Varus, a former governor of Syria who had married into Augustus's family. The appointment of Varus, described as 'a man of a mild character and of quiet disposition, somewhat slow in mind as he was in body, and more accustomed to the leisure of the camp than to actual service in war', might in retrospect be regarded as extremely unwise, but the tribes seemed quiet enough and some had even begun to adopt Roman customs.

There were, however, Germans who detested Roman rule and they were only too eager to take advantage of any opportunity offered by Varus's self-indulgent lifestyle which, through example, had spread downwards and begun to affect the garrison. Prominent among these was Arminius, a young nobleman of the Cherusci, who had seen active service with the Roman Army, and it was he who provided the necessary leadership. Arminius's hatred of Rome was genuine enough, but he also had a personal insult to avenge, for his pro-Roman uncle, Segestes, had pointedly refused him his daughter's hand in marriage, a problem which he solved by eloping with the lady. Although Arminius planned his rising with infinite care both as to the time and place, his intentions could hardly be kept secret and Varus was given ample warning by Segestes. The governor, however, believed that the accusation of treason was merely an extension of the family quarrel and he did nothing. In the autumn of AD 9, the garrison of Germany consisted of five legions, of which three, XVII, XVIII and XIX, about 20,000 men in all, were with Varus in the Minden area. Contrary to their normally strict practice, the legions were accompanied by some 10,000 camp followers, including their families, and a long train of baggage wagons. In September or October Varus was preparing to march back to his winter quarters at Aliso (Haltern) on the Lippe when he received word of an apparently minor rising near the Weser. This was intended by Arminius to draw him off his route and into the difficult country of the Teutoburger Wald, where the main rebel force would attack the column. The area has been described as 'a table-land intersected by numerous deep and narrow valleys, which in some places form small plains, surrounded by steep hills and rocks and only accessible by narrow defiles. All the valleys are traversed by rapid streams, shallow in the dry season, but subject to sudden swellings in autumn and winter. The vast forests which cover the summits and slopes of the hills consist chiefly of oak; there is little underwood, and both men and horse would move with ease in the forests if the ground were not broken by gullies, or rendered impractical by fallen trees.'

Varus not only chose to ignore further warnings from friendly Germans but also complicated his own position when, instead of detaching his wagons and the camp followers and sending them by the direct route to Aliso, he decided to keep them with him. Arminius and his auxiliaries, still above suspicion, actually accompanied the Roman Army during the first phase of its rain-soaked march, and their disappearance during the night was the first hint that something was wrong. Once within the forest the legionaries found the going so difficult that they were forced to construct a track through the morass. At this point the Germans attacked from all sides, cutting the column into sections and slaughtering the camp followers. After the Romans had burned their wagons they made better progress but they found the defiles blocked with fallen trees. The legionary cavalry tried to cut its way out but was killed to a man. Varus and senior officers committed suicide. The agony was prolonged for several days before the last survivors of the dwindling column were overwhelmed. Those captured were nailed to trees, buried alive or ritually sacrificed.

When news of the disaster, which is generally thought to have occurred in the region of Detmold, reached Rome, it is said that Augustus let his hair and beard grow for several months and that he repeatedly beat his head against the wall, bewailing the loss of his legions. During the next five years punitive expeditions were mounted under Tiberius and Germanicus and while these succeeded in inflicting heavy loss, recovering several standards and capturing Arminius's wife, no further attempt was made to bring Germany within the Roman orbit.

Both Sir Edward Creasy and Major-General J. F. C. Fuller regard the defeat of Varus as being one of the decisive battles of the world, commenting that if Arminius had failed the subsequent history of Germany would have been radically altered by prolonged contact with Graeco-Roman influences and that the Saxon emigration to Britain would not have taken place. The battle in the Teutoburger Wald also represents the first important milestone in the history of forest fighting, since it demonstrates not merely the creation of conditions in which a major ambush could be successfully staged, but also the technique by which a large enemy force could be lured onto unfavourable ground in pursuit of a comparatively minor objective and then destroyed piecemeal.

Yet more than seventeen centuries were to pass before the subject would be regarded as worthy of specialist consideration. In North America the Seven Years War, known there as the French and Indian War, was largely fought out in the vastness of the as-yet uncleared wilderness. The French were much better at attracting the support of the Indians to their cause than were the British, and this was to have far-reaching consequences. On 9 July 1755 General Edward Braddock was advancing to attack Fort Duquesne with 1,400 regulars and 450 colonial volunteers when he was ambushed by a force of 900 Indians under French officers on a track near the Monongahela River. The Indians, invisible among the trees, fired at will into the rigid British line. The latter returned useless volleys until Braddock was killed and half his men were down, then the survivors fled; they were rallied and led back to Virginia by a colonial officer, Colonel George Washington.

The major part of Braddock's command consisted of the 44th and 48th Regiments, which had only recently arrived in America and contained a high proportion of raw recruits. In the circumstances, therefore, these men stood their ground far longer than anyone might have expected, but the fact remained that the tactics of the European battlefield were quite unsuited to close country. It was decided to fight fire with fire and shortly after the Monongahela disaster the 60th Royal American Regiment, four battalions strong, was raised at Governor's Island, New York, with the object of 'combining the qualities of the scout with the discipline of the trained soldier'. The new regiment's commanding officer was a Swiss, Colonel Henry Bouquet, and initially most of the men were either German immigrants or Germans recruited in Europe, most of whom would have had some experience of hunting and shooting in their day-to-day lives. They were drilled in open order, both in quick time and double time, taught to load and fire quickly in the standing, kneeling or lying position, instructed in swimming, survival, self-sufficiency and elementary field fortification, and generally required to use their personal initiative. Their training included a period of several weeks spent in the woods during which, apart from a small ration of flour, they relied entirely on whatever game and fish they could shoot or catch. The regiment's success in action earned it a permanent place in the British Army, where it subsequently became known as the King's Royal Rifle Corps, and it now forms part of the Royal Green Jackets.

Simultaneously, an irregular unit was raised from colonists in New Hampshire under the command of Captain Robert Rogers, a native of Massachusetts. Rogers, tall and extremely fit, had lived the life of a frontiersman in his youth and, in addition to his ingrained knowledge of wood craft, he understood Indians and their ways. Rogers'

Rangers were specialists in the art of providing advance and rear guards, intelligence gathering, deep penetration patrols, raiding and sabotage, and they earned themselves a tremendous reputation. On one occasion they marched 50 miles on four feet of snow and fought two successful actions, all in the space of a single day. On another, they marched through a foot-deep swamp for nine days to punish an Indian atrocity. The unit was disbanded in 1763 but it is hardly surprising that when the United States Army decided to form its own commando units during the Second World War, it chose to call them Rangers.

Rogers wrote a set of Standing Orders for his men and, since they are as relevant today for special forces and forest fighting as they were then, they are worth quoting in full:

1. Don't forget nothing.
2. Have your musket as clean as a whistle, hatchet scoured, sixty pounds of powder and ball, and be ready to march at a minute's notice.
3. When you're on the march, act the way you would if you was sneaking up on a deer. See the enemy first.
4. Tell the truth about what you see and what you do. There is an army depending on you for correct information. You can lie all you please when you tell other folks about the Rangers, but don't never lie to a Ranger or Officer.
5. Don't never take a chance you don't have to.
6. When we're on the march we march single file, far enough apart so one shot can't go through two men.
7. If we strike swamps, or soft ground, we spread out abreast, so it's hard to track us.
8. When we march, we keep moving till dark, so as to give the enemy the least possible chance at us.
9. When we camp, half the party stays awake while the other half sleeps.
10. If we take prisoners, we keep 'em separate till we have time to examine them, so they can't cook up a story between 'em.
11. Don't ever march home the same way. Take

a different route so you won't be ambushed.
12. No matter whether we travel in big parties
or little ones, each party has to keep a scout 20
yards ahead, 20 yards on each flank and 20
yards in the rear, so the main body can't be
surprised and wiped out.
13. Every night you'll be told where to meet if
surrounded by a superior force.
14. Don't sit down to eat without posting
sentries.
15. Don't sleep beyond dawn. Dawn's when
the French and Indians attack.
16. Don't cross a river by a regular ford.
17. If somebody's trailing you, make a circle,
come back onto your own tracks, and ambush
the folks that aim to ambush you.
18. Don't stand up when the enemy's coming
against you. Kneel down, lie down, hide
behind a tree.
19. Let the enemy come till he's almost close
enough to touch. Then let him have it and
jump out and finish him up with your hatchet.

The obvious success of Rogers' and
Bouquet's methods led to the widespread
adoption of light infantry tactics by British
regiments in America and with this came a
style of dress more suited to the environment. The brim of the tricorne hat was let
down, the coat was shortened by removing
its elegant turn-back skirts, and the long
gaiters were reduced to short leggings. Hair
was cut short and instead of tramping
through the forest with a napsack full of
pipeclay and hair-dressing, the soldier now
carried extra ammunition and rations. A
hatchet was added to his equipment and the
barrel of his musket was browned to
eliminate reflected light. Fighting in close
country took place in open order, while in
more open terrain a two-deep firing line
replaced the three-deep formation used in
Europe.

Unfortunately, after the war ended in
1763 most of this was quickly forgotten as the
Army reverted to its peacetime routine of
formal drill and spit-and-polish. Indeed,
once the French influence had been removed

from North America, the techniques of
forest fighting seemed irrelevant. However,
on the outbreak of the American War of
Independence in 1775 those techniques had
to be quickly re-learned, and by both sides.
The British won most of the battles but lost
the critical engagements; for their part, the
Americans only began to make real headway
when they ceased relying on locally raised
militias and established a disciplined regular
army.

It was in the West Indies that the British
Army had its first taste of jungle warfare
proper. The produce of these islands, especially sugar, formed a vital element in the
economy of whichever nation owned them.
The islands themselves were rife with tropical diseases, which were fatal to Europeans,
and for this reason the plantations were
worked by slaves transported from West
Africa. It was, however, necessary to provide
garrisons both as a precaution against slave
rebellions and to guard against the ambitions of other nations. At this period a posting to the West Indies was regarded as being
tantamount to a death sentence, for even
when an epidemic was not raging thousands
died from disease every year, the greatest
killer being the grimly named Yellow Jack or
yellow fever. The more thoughtful garrison
commanders did what they could to safeguard their men by sending them to sea in
rotation for brief cruises with the Royal
Navy, but the scourge remained. It was
noticed that those few who actually survived
the disease were seldom affected again and,
because of their immunity, these men often
chose to transfer to incoming regiments
when their own returned home, for life in the
islands could be both pleasant and profit-
able.

The French Revolution, however, shook
the old order in the West Indies to its foundations. On the French islands the slaves
followed the lead of the Paris mob in pro-
claiming Liberty, Equality and Brotherhood

for all and rose, slaughtering their masters. Simultaneously, French agents efficiently destabilised British possessions until it became necessary to despatch troops to restore order. Most of these were Light Dragoon regiments, which combined the mobility of cavalry with the flexibility and initiative of light infantry; one, the 20th Light Dragoons, was raised at the expense of the Jamaica government for permanent service in the island and, as a concession to the climate, they wore a lightweight helmet made from tin instead of the usual fur-crested leather Tarleton helmet.

During this prolonged counter-insurgency operation the enemy would regularly mount ambushes and then, having emptied a few saddles, retire into the bush, but as their confidence grew they rashly attempted open confrontations with the troops and suffered severely. The final phase consisted of tracking the insurgents to their strongholds, which were usually situated in the jungle-covered mountainous interior, and during this the advance was usually led by a troop of picked marksmen; in some instances small howitzers were manhandled along the trails and used to shell the rebels out of their positions. By the end of the 1790s the situation had been brought under control and regiments began to return home. As always, Yellow Jack had made it quite clear where his sympathies lay and had inflicted far heavier losses than the insurgents. The 17th Light Dragoons recorded that in a seven-month period of 1796 the five troops serving in San Domingo lost 7 of their 12 sergeants, 76 of their 116 privates and both trumpeters; of the 400 men of the 14th Light Dragoons who had sailed for the West Indies in 1795 only 25 returned two years later, and although some may have volunteered to remain behind, the comparable experience of other regiments suggests that as many as 300 may have died. Happily, the West Indies had all but seen the last of slavery, for in 1807 it was abolished by the British government.

By the nineteenth century, most European powers had acquired coastal trading stations in tropical lands, usually with the permission of friendly local tribes. From these simple beginnings a pattern evolved. The local tribe would be attacked by its more powerful neighbours and ask the Europeans for help. The Europeans would respond but the small forces immediately available would sustain a reverse. It then became necessary to despatch a much larger force, which defeated the common enemy, whose territory was annexed. In this way the European influence was extended inland until the small trading station became the centre of a colony, the British fighting several such wars in West Africa and Burma while the French did likewise in Dahomey and Indo-China. The whole process was accelerated sharply by the industrialisation of the West, since such commodities as rubber, oil, tin and copper were to be found in these regions and their possession became essential, not merely for commercial reasons, but also because without them it was impossible to wage a modern war.

Nevertheless, the techniques of jungle fighting attracted little attention from the world's major armies, although at the turn of the century the then Colonel C. E. Callwell produced his famous analysis of colonial campaigns, *Small Wars*, in which he included a chapter on the subject. Bush warfare, he pointed out, was essentially an affair of surprises and ambuscades in which the enemy would instinctively attack the flanks and rear of an advancing force. For this reason, troops should advance with strong flank and rear guards, the effect being that of a broad flexible square. Progress would inevitably be slow because of the need to hack a way forward through the thick undergrowth, and there would be times when officers would have to use their compasses to maintain direction because of the restricted view

Gurkha infantry storm a village during the Third Burma War, 1885–86 (National Army Museum)

Dacoit ambush, Third Burma War. The troops' reaction is to respond with rapid fire and search for the enemy's flank. (National Army Museum)

imposed by vegetation. Since much of the fighting would take place at close range, good junior leadership and local control were essential. If the troops were fired upon, they

'should promptly charge towards the spot whence the fire comes. To stand still is the worst thing they can do. Some difference of opinion exists as to whether a volley should precede the charge, although officers experienced in bush warfare generally favour this procedure because it is often the only way of causing casualties among the enemy. If the troops are armed with magazine rifles, magazine fire for a few seconds to start with would generally seem the wisest plan. But above all things the enemy must not be allowed time to fire a second volley, and to prevent this a prompt bayonet charge is almost always expedient.'

An alternative reaction to this was recommended by General Sir Frederick Roberts during the Third Burma War of 1885-86:

Supposing, for instance, the fire of the enemy to be delivered from the right, a portion of the force should be ready to dash along the road for 100 yards or so, or until some opening in the jungle offers itself. The party should then turn to the right and sweep round with a view to intercepting the enemy in his flight. A party in rear should similarly enter the jungle to the right with the same object. The centre of the column would protect the baggage or any wounded men. The different parties must be previously told off, put under the command of selected leaders, and must act with great promptitude and dash.

If the enemy's presence was suspected, Callwell was in favour of raking the undergrowth with precautionary volleys before he could spring his ambush. He also emphasised the importance of recruiting scouts from among the local population:

Scouting in the bush is an art in itself, an art which can only be learnt by experience and a duty for which all are not fitted. Natives of jungle-grown countries have a natural talent for wood craft, for detecting footmarks and for noting details not observable to the un-

initiated eye. Scouting is best left to irregulars enlisted on the spot.

Baggage and supply trains presented special problems. On balance, Callwell preferred pack animals to native porters since the latter were prone to throw down their loads and bolt if the situation became difficult. He recommended either leaving the supply column behind under escort and only bringing it forward when the way ahead had been cleared for several miles, or splitting it into sections which were disposed among the main force. It was particularly important that the supply train should not outstrip the flanking columns which formed the sides of the flexible square as these would inevitably make slow progress as they cut their way forward. The one subject upon which Callwell admits that he has very little to offer is conducting a retreat, the reason being that the campaigns of the period were won without undue difficulty and without encountering a first-class, fully-equipped enemy.

Of these small Victorian jungle wars, perhaps the most interesting and instructive is the Ashanti campaign of 1874, fought in the Gold Coast, known today as Ghana. The Ashanti tribal confederation, ruled by King Coffee Calcalli, were troublesome neighbours given to demanding tribute and taking hostages, and in 1873 their incursions reached such serious proportions that it was decided to mount a punitive expedition under General Sir Garnet Wolseley. The War Office had serious reservations about committing British troops to this notoriously fever-laden coast but at length despatched the Royal Welch Fusiliers, the Black Watch and the Rifle Brigade; significantly, they remained at sea until Wolseley was ready to commence his 100-mile march inland on Kumasi, the enemy's capital, in January 1874. In addition, Wolseley had available the 1st and 2nd West Indian Regiments, two locally raised native regiments, and a naval brigade consisting of sailors and marines;

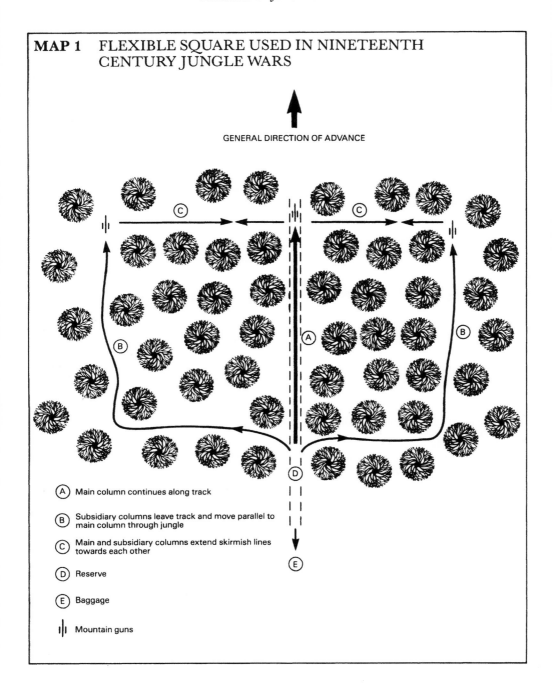

MAP 1   FLEXIBLE SQUARE USED IN NINETEENTH CENTURY JUNGLE WARS

GENERAL DIRECTION OF ADVANCE

(A) Main column continues along track

(B) Subsidiary columns leave track and move parallel to main column through jungle

(C) Main and subsidiary columns extend skirmish lines towards each other

(D) Reserve

(E) Baggage

⫞ Mountain guns

because of the obvious impracticality of employing field guns in the bush, artillery support consisted of a battery of mountain guns and rocket tubes under a Captain Rait. Leaving the West Indians and part of the native infantry to guard his line of communications, Wolseley advanced inland against minor opposition until, early on 31 January, he encountered the main body of the Ashanti Army near the village of Amoaful. The course of the action is described by the noted contemporary novelist and military historian, G. A. Henty:

> The plan of operations had already been determined upon. The Black Watch were to form the main attacking force. They were first to drive the enemy's scouts from the little village of Agamassie and were then to move straight on, extending to the left and right of the path, and, if possible, to advance in a skirmishing line through the bush. Two guns of Rait's battery were to be in their centre, and to move upon the path itself. Half the naval brigade and Wood's (native) regiment were first to cut a path to the right and then turn parallel with the main path, so that the head of the column should touch the right of the skirmishing line of the Black Watch, while the other half of the naval brigade, with Russell's (native) regiment, was to proceed in a similar fashion on the left. Two companies of Fusiliers were to come on behind the headquarters staff; the Rifle Brigade were to remain in reserve. The intention was that the whole should form a sort of hollow square, the columns on the left and right protecting the Highlanders from the flanking movements upon which the Ashanti were always accustomed to rely for victory. With each of the flanking columns were detachments of Rait's battery with rocket tubes.

The hamlet was taken without difficulty and the Black Watch pressed on into the bush beyond while the flanking columns diverged to the left and right. Shortly after, the outbreak of heavy firing indicated that all three elements of Wolseley's attack force was in contact with the main body of the Ashanti Army.

The scene bore little resemblance to that presented by any modern battlefield. The Ashanti bush consists of a thick wood of trees some forty or fifty feet high, covered and interlaced with vines and creepers, while the heat and moisture enable a dense undergrowth to flourish beneath their shade. Above all tower the giants of the forest, principally cotton trees, which often attain a height of from 250 to 300 feet. Progress through this mass of jungle and thorn is impossible even for natives, except where paths are cut with hatchet or sword. The ground across which the Highlanders were trying to force their way was more open than usual, owing to the undergrowth having been cleared away to furnish fire for the little village. It was somewhat undulating and the depressions were soft and swampy. Each little rise was held obstinately by the enemy, who, lying down behind the crest, behind trees, or in clumps of bush, kept up an incessant fire against the Black Watch; and even the aid of Rait's two little guns and two rocket troughs failed to overcome their resistance. The two flanking columns encountered even more strenuous opposition: before they could advance into the bush a way had to be cut for them by the natives under the orders of the Engineer officers. Although the troops endeavoured to cover this operation by incessant fire into the bush on either side, the service was a desperate one. Several of the men fell dead from the fire of their hidden foes, others staggered back badly wounded, and Captain Buckle, of the Royal Engineers, one of the most zealous and energetic officers of the expedition, fell mortally wounded by two slugs in the neighbourhood of the heart.

Little wonder was it that, although the natives behaved with singular courage, at times they quailed under the fire to which they were exposed; consequently the advance of the two [flanking] columns soon came to a standstill, and the men lying down kept up a constant fire on the unseen enemy, directing their aim solely at the puffs of smoke spurting from the bushes. So difficult was it to keep direction in this bush that both columns had swerved from the line on which it was intended that they should advance. The roar of fire was so general and continuous that none of the three columns were in any degree certain as to the direction in which the others lay, and from each of them messenger after mes-

Above *Native troops beat off an attack on their village during the Second Ashanti War, 1873–74.* (National Army Museum)

Below *Light artillery and infantry in action at Amoaful, 31 January 1874.* (National Army Museum)

senger was sent back to Sir Garnet Wolseley, who had taken up his position with his staff at the hamlet, complaining that the men were exposed to the fire from the other columns.

The noise was, indeed, out of all proportion to the number of combatants. The Ashantis use enormous charges of powder—which, indeed, would be absolutely destructive to the old Tower muskets with which they were armed were these loaded with tightly-fitting bullets. This, however, was not the case, as on the powder three or four slugs of roughly chopped-up lead were dropped loosely down: the noise made by the explosion of the muskets so charged was almost as loud as that of small field pieces; and, indeed although but two or three hundred yards from the village the reports of Rait's mountain guns were absolutely indistinguishable in the din. Well it was for our soldiers that the enemy used such heavy charges, for these caused the muskets to throw high, and the slugs for the most part whistled harmlessly over the heads of the troops and almost covered them with showers of leaves cut from the trees overhead.

For an hour this state of things continued, then two companies of Fusiliers were ordered to advance along the main path and assist the Black Watch in clearing the bush, where the Ashantis still fought stubbornly not two hundred yards from the village. Two companies of the Rifle Brigade were sent up the left-hand road to keep the path intact up to the rear of the naval brigade, while, on the right, the rear of Colonel Wood's column was ordered to advance further to the right, so that the column might form a diagonal line, and firing to their right only, not only cover the flank of the Black Watch, but do away with the risk of stray shots striking them. Wounded were now coming fast into the village—Highlanders, Rifles, naval brigade and natives.

On the left the firing gradually ceased, and Colonel McLeod, who commanded there, sent in to the general to say that he was no longer hotly attacked, but that he had altogether lost touch with the left of the Black Watch. He was therefore ordered to cut a road north-east until he came in contact with them. He experienced resolute opposition, but the rockets gradually drove the Ashantis back. In the meantime, the Highlanders were fighting hard. In front of them was a swamp, and on the rise opposite the ground was covered with the little arbours that constitute an Ashanti camp. Not an enemy was to be seen, but from the opposite side the puffs of smoke came thick and fast, and a perfect rain of slugs swept over the ground. The path was so narrow that Rait could bring but one gun into position. This he pushed boldly forward, and, aided by Lieutenant Saunders, poured round after round of grape into the enemy until their fire slackened and the Black Watch were again able to advance.

At noon, five hours after the battle had begun, the Highlanders broke out of the bush into the cleared area surrounding Amoaful.

For a short time the Ashantis kept up a fire from the houses and from the other end of the cleared space, but the Black Watch soon drove them from the houses; and a shell from a gun fell among a group at the farther end of the clearing and killed eight of them, and the rest retreated at once.

The day's fighting, however, was not quite over. A determined counter-attack on the right-hand column came close to isolating the Black Watch until it was beaten off, and another group of Ashanti assaulted the previous night's camp at Quarman. This was held by some 40 men of the 2nd West Indian Regiment and half a company of native infantry, who not only held their assailants at bay but also sallied forth to bring in a convoy; when a company of the Rifle Brigade was sent back to reinforce the garrison the Ashanti melted into the bush.

At Amoaful the Ashanti had between 800 and 1,200 killed and about the same number wounded. Wolseley's losses amounted to a mere 4 killed and 194 wounded, rather more than half being sustained by the Black Watch. Although he had only four days' rations in hand and it would be almost a week before more became available, Wolseley decided to advance at once on Kumasi, some 20 miles distant. He reached the enemy's capital on 4 February to find that its entire population of 30,000 had fled. Near the great fetish tree several thousand skeletons and decomposing bodies provided ample evidence of the ritual sacrifice which had earned the place its evil reputation, and further grisly proof was found in King Coffee's palace itself. Wolseley was already concerned that the climate was beginning to affect his troops and, when torrential rain signalled the start of the wet season, the following day he began his return march to the coast, having first blown up the palace and burned the city to the ground. On 13 February King Coffee signed a treaty of submission and paid a modest indemnity.

By the time that the British contingent had re-embarked aboard its troopships signs of fever were already apparent; the Governor appointed by Wolseley in March died from its effects before April was out. Wolseley returned home to receive the personal thanks of Queen Victoria and of Parliament, and to enjoy a hero's welcome. The Ashanti expedition had indeed been well-planned and capably executed, so much so that 'All Sir Garnet' came to be used as a term for efficiency.

Wolseley was to render his country much greater services (see *Desert Warfare* in this series) and was the prototype of Gilbert and Sullivan's Very Model of a Modern Major-General.

In 1895 widespread Cuban dissatisfaction with Spanish colonial rule erupted into full-scale rebellion. The Spanish Army was fundamentally unsuited to the type of guerrilla war imposed by the insurgents, suffered from disease as much as any other, and received neither intelligence nor any other form of assistance from the implacably hostile population. A British cavalry officer who was present in Cuba at the time noted the malaise which affected the Spaniards.

There was a complete absence of any general plan. Columns moved about haphazard in the woods, fighting the enemy where they found them and returning with their wounded to the towns when they were weary of wandering. Their method of warfare was essentially defensive. They held great numbers of towns and villages with strong garrisons. They defended, or tried to defend, long lines of communications with a multitude of small blockhouses. They tried to treat the rebels as though they were merely agrarian rioters, and to subdue the revolt by quartering troops all over the country. The movement was on a scale far exceeding the scope of such remedies; it was a war, and this the Spanish Government would never recognize. Over all the petty incidents of guerrilla skirmishing, the frequent executions and the stern reprisals threw a darker shade.

The rebels were strongest in the east but their hold on the country quickly spread westwards until Havana itself was threatened. In 1896, however, General Valeriano Weyler y Nicolau was appointed commander of the Spanish forces and the war began to assume a definite pattern. Weyler began by dividing the island into tactical zones, using

Men of the US 4th Infantry Regiment form a firing line for the benefit of the camera. Luzon, Phillipine Islands, 1900. (US Army Military History Institute)

lines of barbed wire entanglements and blockhouses to restrict the movement of the *insurrectos*. He then deprived the insurgents of local support by placing the civilian population of the worst affected areas in 'concentration camps'. The rebels soon began to lose ground and were again confined to the mountains of the east. Unfortunately, while the situation had been brought under control in the purely military sense, conditions within the camps were so bad that thousands died from lack of adequate food and epidemics aggravated by overcrowding.

This aspect of the war, together with tales of real or imaginary Spanish atrocities, was widely reported in the American press and generated such indignation that the United States government demanded Weyler's recall. Anxious to reduce tension, the Spanish obliged in October 1897, but neither this, nor the ending of the camp system, nor the offer of home rule to the Cubans, received much attention. Together,

politicians and press had so thoroughly stirred up the American public that the mysterious destruction of the battleship USS *Maine* in Havana harbour on 15 February 1898 made war with Spain inevitable.

Following the destruction of her Cuban and Philippine naval squadrons, Spain sued for peace. Under the terms of the Treaty of Paris, she relinquished her sovereignty over Cuba and sold the Philippine Islands to the United States for the sum of $20 million. Almost immediately, the Americans found themselves fighting a jungle war of their own.

Rudyard Kipling, then resident in the United States, might celebrate the emergence of the nation as a colonial power with his famous exhortation to 'take up the white man's burden', but the fact was that the 'new-caught sullen peoples' of the Philippines objected strongly to the idea. Led by Emilio Aguinaldo, the Filipinos had sought independence from Spain and, after they

Filipino insurrectos captured at the Battle of Rio Grande. (USAMHI)

Left *Rebel trenches near Rio Grande bridge. Although all but forgotten now, the war in the Philippines found almost as little favour among the American people as did the war in Vietnam some 60 years later — for much the same reasons.* (USAMHI)

Below *An American patrol escorts prisoners captured at Porac.* (USAMHI)

A 2.75-inch mountain gun in action in the German Cameroons, 1915. The weapon is breech-loading and is equipped with recoil buffers but is much the same size as those used at Amoaful. (Imperial War Museum)

had assisted American troops to capture Manila, they felt that they deserved better than a simple change of masters. They blockaded the American garrison in Manila but were driven off when fighting broke out on 4 February 1899. A protracted guerrilla campaign ensued in which the American commander, General Arthur MacArthur, the father of General Douglas MacArthur, was forced to mount counter-insurgency operations in Luzon, Mindanao, Sulu and the Visayas. Aguinaldo was captured on 23 March 1901 and issued a proclamation calling for peace, although another year was to pass before the last of the Filipino leaders surrendered; even then, the Moro tribesmen in the southern islands remained in a state of unrest and were not finally subdued until 1905.

Altogether, 100,000 American troops were employed in putting down the Philippine Insurrection; of these, 4,243 were killed and 2,818 were wounded in action. Some 20,000 rebel Filipinos were killed and approximately 200,000 civilians died as a result of famine and disease. In the United States itself the war, which cost $600 million, was almost as unpopular as that fought in Vietnam 60 years later, and for much the same reasons.

The first campaigns in which regular troops faced each other in a jungle environment took place during the early years of the First World War and have now been largely forgotten. Imperial Germany had joined the colonial powers somewhat late but her tropical possessions included an enclave in north-east New Guinea, the Bismarck Archipelago, the Caroline, Marshall and Marianna Islands, Samoa, Bougainville and, on the west coast of Africa, Togoland and the Cameroons. The majority were under-developed, poorly settled and, when war broke out, completely isolated from the homeland. By October 1914 all of Germany's Pacific territories were in Allied hands, having either surrendered on demand or after a short token resistance.

Togoland, squeezed between British Gold Coast and French Dahomey, also fell after a

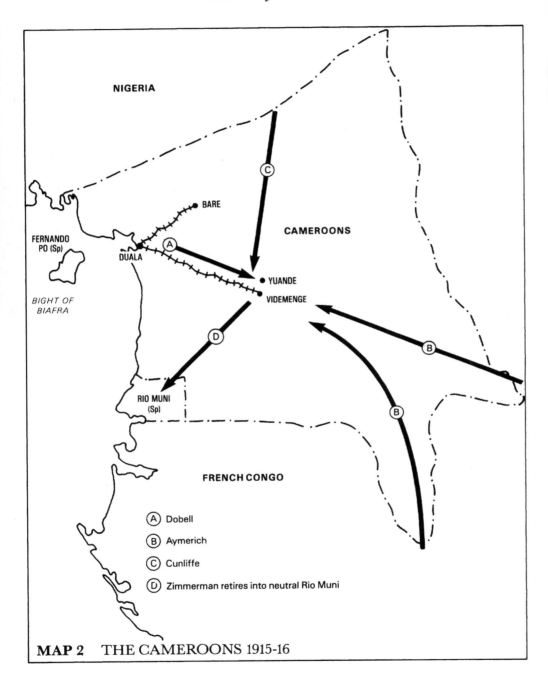

NIGERIA

BARE

CAMEROONS

FERNANDO
PO (Sp)

DUALA

BIGHT OF
BIAFRA

YUANDE

VIDEMENGE

RIO MUNI
(Sp)

FRENCH CONGO

Ⓐ Dobell

Ⓑ Aymerich

Ⓒ Cunliffe

Ⓓ Zimmerman retires into neutral Rio Muni

MAP 2   THE CAMEROONS 1915-16

24

French Senegalese troops on the march near Yuande, German Cameroons. (IWM)

brief campaign. The principal port, Lome, surrendered to a British cruiser and the German garrison, consisting of 3,000 native troops under German officers, retired inland to Atakpame. Simultaneously, the western frontier had been crossed by part of the Gold Coast Regiment, riding in motor transport, and two French columns marched east from Dahomey. Together with the main British column moving north along the railway from Lome, 558-strong with 2,000 porters, these troops converged on Atakpame. The Germans made a stand at the River Khra on 22 August, using their machine guns to good effect, but surrendered five days later.

The Cameroons, located between Nigeria and French Equatorial Africa and one-third larger than Germany itself, were also attacked in August 1914, but proved a much tougher nut to crack. The capture of the port of Duala on 27 September effectively isolated the colony and the Germans withdrew into the interior. From Duala the Allies pushed slowly inland along the two railway lines and consolidated their hold on the coastal region.

It was, however, soon apparent that the 5,000-strong Allied force was too small to tackle the vast and difficult hinterland in which the well-equipped German forces commanded by Colonel Zimmerman, originally consisting of 200 German officers, 3,200 native troops, 1,500 armed constabulary and three artillery batteries, were rapidly expanded by local enlistment to a strength of 2,800 Germans and over 20,000 Africans. The Allies were also steadily reinforced and although some of their columns which crossed the frontier from neighbouring territories were sharply checked by

King's African Rifles provide an escort for an Indian mountain battery. German East Africa, September 1917. (IWM)

German counter-attacks, by the onset of the rainy season Zimmerman's command had been whittled away in small actions to 1,300 Germans and 9,000 natives, centred on Yaunde, the summer capital.

On the coastal sector, commanded by Major-General Sir Charles Dobell, the logistic aspects of the campaign were eased by the arrival of a company of Ford motor lorries, each capable of doing the work of 150 porters; also delivered were a number of radio sets and a single Rolls-Royce armoured car.

By the end of September 1915 the Allies were ready to commence their final advance, the objective of which was Yaunde. Dobell, with 729 Europeans and 7,555 Africans, of whom half were French or Belgian, was to continue fighting his way east along the railway while a 7,000-strong French force under General F. Aymerich converged from the opposite direction; simultaneously, a third column under Brigadier-General F. J. Cunliffe was to advance southwards from the area of the Nigerian frontier.

The Germans offered stiff resistance but were pushed steadily back and on 25 December decided to abandon Yaunde. Dobell's leading units entered the town

unopposed on 1 January 1916 and were joined by Cunliffe and Aymerich four days later. Early in February Zimmerman marched his surviving 800 German and 6,000 African troops into the neutral Spanish colony of Rio Muni and was interned. His achievement was not quite comparable to that of General Paul von Lettow-Vorbeck, who waged a successful guerrilla campaign on the East African plateau until November 1918 but, assisted by the climate and the difficult terrain, he had managed to tie down large numbers of Allied troops for 18 months.

Apart from the use of machine guns by both sides and the limited employment of motor transport, the campaigns in Togoland and the Cameroons were little influenced by modern technology and really belonged to the nineteenth century. Greater events elsewhere soon eclipsed whatever lessons there were to be learnt and in the years which followed the First World War the prospect of the Imperial powers becoming involved in further jungle campaigns seemed remote; yet it was to be in jungle conflicts that the foundations of those same empires would be shaken to the point of disintegration.

CHAPTER 2

THE ALL-CONQUERING BICYCLE

Until 1921 the Empire of Japan had enjoyed excellent relations with the West, but in that year a conference was held in Washington the results of which marked the beginning of a process which would turn old friends into bitter enemies. The purpose of the conference was to agree limitations on the number of battleships and aircraft carriers which each of the major naval powers could possess and thereby prevent the crippling expense of a naval construction race. Ultimately, it was decided that for every five capital ships maintained by the Royal Navy and the United States Navy, Japan could maintain three. Since the Imperial Navy was a source of great pride to the Japanese such an agreement involved serious loss of face. In itself, this might have been considered tolerable had not the United Kingdom, for the sake of balance, decided to terminate its long-standing treaty of alliance with Japan at the same conference. To the Japanese, such an act intentionally diminished their international standing and was not to be forgiven. Their isolation was further confirmed in 1924 by the United States' refusal to permit further Japanese immigration.

The Great Depression, coupled with tariff barriers erected by the West as a defence against cheap imports, created havoc in Japanese industry and led to widespread hardship. By the early 1930s the armed services had become the dominant force in Japanese politics and they were

determined that Manchuria would be preserved as an outlet for Japan's goods, despite the open enmity of Chiang Kai-shek, who had emerged as China's leading warlord in 1928. An incident was provoked on 18 September 1931 which resulted in Japanese troops overrunning the whole of Manchuria and establishing the puppet state of Manchukuo. Next, Japan withdrew from the League of Nations and in 1934, having renounced the restrictions imposed by the Washington Naval Treaty and those which were extended to other classes of warship by the London Naval Conference of 1930, embarked on an impressive naval construction programme.

On 7 July 1937 outright war broke out between Japan and China. Initially, the Japanese were victorious, but the Chinese withdrew into their vast hinterland and continued to resist stubbornly from there. The strategic aim of Japan, therefore, became the complete isolation of China from foreign aid. In 1940, taking advantage of British and French weakness following German victories in Western Europe, she forced the temporary closure of the Burma Road and compelled the French Vichy government to grant bases in Indo-China.

The savagery with which Japan had waged her war in China had, however, provoked a feeling of deep repugnance in the West. In July 1940 President Roosevelt banned the export of strategic materials to

Japan and a year later froze all Japanese assets in the United States. The British and Dutch governments did likewise, effectively preventing the Japanese from trading for the supplies of oil, tin and rubber which were to be found in their Far-Eastern possessions. The Japanese found themselves faced with a choice: they could either abandon their campaign in China, or they could seize by force the raw materials which were essential for the survival of their war machine. It was inevitable that they should choose the latter, although those in power were fully aware that Japan could not sustain a prolonged conflict on several fronts. Thus, their attack on the American Pacific Fleet at Pearl Harbor and the invasion of British, Dutch and American territories were not seen by the Japanese as treacherous acts, but as courageous pre-emptive strikes in what was intended to be a short war which would leave Japan in a strong negotiating position. Again, with the United Kingdom deeply committed to its war against Germany and Italy, Holland under German occupation, the United States still unprepared for war

and the ancient enemy, Russia, effectively neutralised by Hitler's invasion, the timing of such strikes would never be better.

Unfortunately for the Western Allies, they seriously underestimated Japan's capacity to fight a modern war, particularly in the air. The Imperial Navy had initially been trained by the British and was extremely efficient. It possessed excellent night fighting techniques and had gone to infinite pains to perfect the devastating oxygen-fuelled Long Lance torpedo, one of the most effective weapon systems of the Second World War. For the moment, it also possessed the finest naval air arm of any of the combatant nations, requiring the highest standards of performance from its aircrew candidates, and its Mitsubishi A6M Zero carrier fighter was faster and more manoeuvrable than anything which the Allies had immediately available, although it was very vulnerable to return fire.

The Japanese Army was an infantry

Japanese 75mm mountain artillery battery in action. (IWM)

army which favoured the storm troop tactics employed during the German offensives of 1918. Heavy emphasis was placed on maintaining the momentum of the advance and if opposition was encountered the attackers flowed round it until it was eliminated by follow-up troops. The infantryman's small arms and automatic weapons were comparable to those of other armies, and extensive use was made of mortars. The field artillery was armed with 75 mm and 105 mm weapons which were lighter and possessed longer range than their Western counterparts, but the army's 37 mm anti-tank guns were hopelessly obsolete and their replacement, the 47 mm Model 01, introduced in 1941, was only effective at comparatively short range. Armoured vehicles in service included a variety of two-man tankettes, which were little more than tracked machine gun carriers, the Type 95 light tank, the elderly Type 89 medium tank and its successor, the Type 97. All were cramped, thinly armoured and under-gunned, but despite being the worst tanks in the world were better than no tanks at all. Armoured units were employed in support of infantry operations and had no experience at all of fighting a tank battle. Being a largely unmechanized army, the Japanese logistic system was primitive and relied to a great extent on local resources, including animal transport, conscripted labour and bicycle trains. While the troops were advancing they could live off the country and captured enemy rations, leaving the transport services free to concentrate on bringing up ammunition and essential stores; if, on the other hand, the advance was checked, serious shortages quickly developed. Nor was the army's communications apparatus suited to a modern war, with the result that co-ordinated action by separate formations was extremely difficult, especially in jungle terrain.

Despite its various shortcomings, there were two reasons why the Japanese Army of 1941 was so formidable. First, it was a very experienced army, battle-hardened by years of active service in China. Secondly, in addition to the iron discipline it imposed, it possessed a profound religious conviction, ingrained from birth, that to die in the Emperor's cause was the noblest of ideals. Indeed, even before he left his homeland, the Japanese soldier believed that he had already given his life to the Emperor, and would sometimes burn nail and hair clippings in a ritual funeral ceremony. All armies talk of fighting to the last man, but only the Japanese did so as a matter of course. In such an army wounds and sickness had to be endured until a man could no longer fight. The ultimate disgrace was to fall alive into the enemy's hands; suicide was infinitely preferable, and if men were too badly injured to perform this act for themselves, their retreating comrades would shoot them as a final act of mercy. Because the concept of surrender was unintelligible to them, the Japanese regarded their Allied prisoners as men totally lacking in honour and treated them accordingly, although this in no way excuses the barbarities that were perpetrated. This same combination of rigid discipline and dedication also produced the two major tactical faults for which the Japanese Army became noted, namely a slavish obedience to orders despite radically altered circumstances, and the repetition of failed attacks over the same ground notwithstanding the heavy casualties incurred.

Because of the dramatic nature of their early victories in Malaya, Burma, the Dutch East Indies and the Philippines, a feeling arose that the Japanese were experts in jungle warfare. The truth was that they had less practical experience of the jungle than their opponents, a fact which caused them such concern that in January 1941 they established a special unit in Formosa to study the problem while details of the enemy topography, bridges, road and track systems were gathered on the ground itself by 'business-men' working under the centralised control of their various military attachés. In the final analysis, their victories stemmed from the application of basic military principles, air power, command of the sea and the will to win. Conversely, they were gained over armies which were essentially flawed, lacked naval and air support, and were psychologically unprepared for the encounter.

Malaya, with its priceless assets of tin and rubber, was high on the list of Japanese priorities. The possibility of a Japanese invasion had been considered by the British as early as 1937, when the then Brigadier Arthur Percival, serving as Chief of Staff to the General Officer Commanding Malaya, Major-General William Dobbie, correctly forecast that this would be made through southern Thailand. Percival's recommendation that the whole of Malaya should be defended was accepted, but his suggestion that the jungle did not present the impenetrable obstacle it was generally thought to be was quietly set aside. In 1940 Dobbie's successor, Major-General Lionel Bond, again emphasised that the whole of Malaya should be held and additionally stressed the importance of air support. As a result of this several new airfields were constructed, but the only fighter aircraft allocated were four squadrons of obsolete Brewster Buffaloes; in the event, the new airfields proved to be more of a liability than an asset, since even if they were non-operational they had to be denied to the enemy and this involved dispersion of troops over a wide area. In other respects, the attitude of service chiefs to the defence of Malaya and Singapore was relaxed and departmental, as was that of the senior civil officials upon whom would fall the burden of establishing emergency services in the event of war.

In general, the good life enjoyed by the colony remained undisturbed despite the threat, with political, economic, commercial and even social considerations taking precedence over the military. In London, Churchill also doubted whether the Japanese would risk outright war, although in May 1941 he was reminded by General Sir John Dill, the Chief of the Imperial General Staff, that Singapore was the lynchpin of the British presence in the Far East and that its security came before that of Egypt. Churchill ignored his advice and continued to reinforce the Middle East, where he had high hopes of a victory over the Axis. The accepted view was that Singapore itself was impregnable to an assault from the sea, and that whatever happened on the Malayan mainland, the arrival of a British fleet within 70 days would restore the situation. Where such a fleet was to come from, with the Royal Navy stretched to its limits in the Mediterranean, the Atlantic and the Arctic, does not seem to have attracted much serious thought; and how it would fare against Japan's modern, superbly trained and equipped carrier squadrons was soon to be tragically demonstrated.

In 1941 the British Commander-in-Chief Far East was Air Chief Marshal Sir Robert Brooke-Popham, under whom Percival, now a lieutenant-general, was appointed GOC land forces in May. Percival had enlisted as a private during World War I, being subsequently commissioned and decorated for gallantry, and had acquired a reputation as an able and perceptive staff officer. He did not share the commonly held view that the Malayan terrain was tank-proof and in August requested two armoured regiments, plus additional anti-tank and anti-aircraft guns. For the moment, his request was denied, with terrible consequences.

Notwithstanding his acute perception, his undoubted personal courage and essential humanity, Percival lacked the drive and charisma needed to inspire his command, which consisted of the 9th Indian Division (Major-General A. E. Barstow), 11th Indian

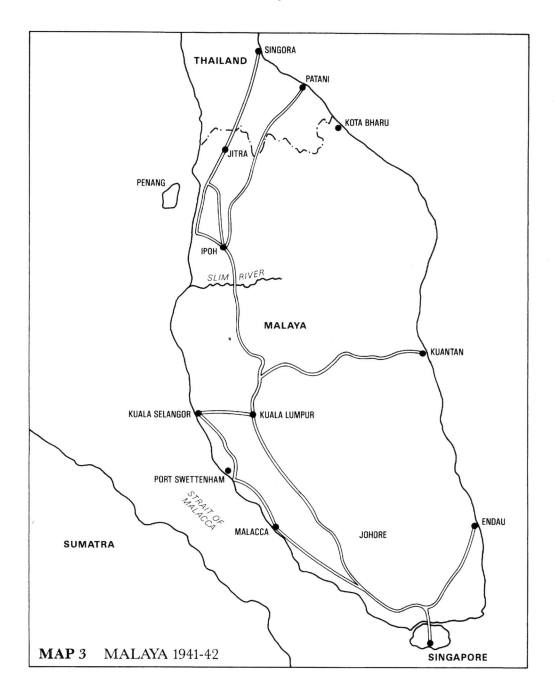

MAP 3   MALAYA 1941-42

Division (Major-General D. M. Murray-Lyon), each of two brigades, the 8th Australian Division (Major-General H. Gordon Bennett), also of two brigades, the independent 12th and 28th Indian Brigades, the 1st and 2nd Malaya Brigades, and a number of independent Indian infantry battalions performing guard duties.

In all, Percival possessed thirty-one infantry battalions, seven field artillery regiments, two anti-tank regiments, five anti-aircraft regiments and eight engineer companies—a total of 80,000 men. Despite this being approximately one-third short of his requirements, it was still an apparently impressive force—that is, until one looked beneath the surface. In general, it was inadequately trained and very few of those present had active service experience. The only armoured vehicles available were tracked weapons carriers and a few armoured cars, the latter being armed with machine guns. The Indian formations were newly raised and contained a high proportion of recently commissioned British officers, many of whom had not yet mastered Urdu, the Indian Army's *lingua franca*, and were consequently unable to communicate with their men. Overall, there was a dangerous tendency to despise the Japanese.

By the autumn of 1941 the prospect of war in the Far East had become a probability and Percival deployed his troops accordingly. The defence of northern Malaya was entrusted to Lieutenant-General Sir Lewis Heath's III Corps, with 9th Indian Division's brigades at Khota Bharu and Kuantan on the east coast, 11th Indian Division near the Thai frontier, covering the better routes south on the west coast, and 28th Indian Brigade in corps reserve at Ipoh. The 8th Australian Division and 12th Indian Brigade were held in army reserve in the south and the two Malaya-brigades were stationed on Singapore Island. On 2 December the battleship *Prince of Wales* and the battle-cruiser *Repulse*, which Churchill hoped would serve as a deterrent, entered Singapore harbour and the British cup of complacency was full. The Japanese were not impressed and two days later their heavily escorted invasion fleet left Hainan.

The officer selected to command the invasion force was Lieutenant-General Tomoyuki Yamashita, who possessed an excellent record and was noted for both his insight and the ability to obtain the last ounce of effort from his men. He had incurred the personal enmity of Prime Minister Tojo and Field Marshal Hisaichi Terauchi, the Commander-in-Chief South-East Asia, and his selection might therefore seem somewhat strange to Western eyes, yet it was typically Japanese. If Yamashita was to fail, and it was by no means certain that he would not, he would be dismissed and publicly disgraced; if, on the other hand, he was to succeed, Tojo and Terauchi would gain the credit of having chosen well.

Yamashita's command was designated the Twenty-Fifth Army and consisted of the experienced 5th and 18th Divisions, commanded respectively by Lieutenant-Generals Takuro Matsui and Renya Mutaguchi, plus Lieutenant-General Takuma Nishimura's Imperial Guards Division, the 3rd Tank Brigade with 80 tanks, two field artillery regiments and a strong engineer contingent. As Percival had predicted, Yamashita intended landing on the Kra Isthmus in Thailand, where the ports of Singora and Patani and their adjacent airfields would prove invaluable during the build-up phase, but he also added the airfield of Khota Bharu to his list of primary objectives.

The landings took place during the early hours of 8 December. Those on the Thai coast were unopposed but at Khota Bharu the 8th Indian Brigade offered fierce resistance, piling up Japanese casualties while RAF Hudsons destroyed one transport

Type 97 medium tank of the 1st Tank Regiment in action in Malaya. The frame antenna indicates that the vehicle is a command tank. (Via Isobe Takuo)

and seriously damaged two more. By evening, however, the landing force had secured a beach-head and the 8th Brigade received permission to withdraw to the south, although this meant abandoning the airfield.

It had been anticipated, correctly, that the Thais would offer only token resistance and for this reason it had been planned to send the major part of 11th Indian Division across the frontier with a view to seizing Singora before the Japanese could establish themselves. Unfortunately, permission to execute this operation, codenamed *Matador*, was denied by the Cabinet on the grounds that it was provocative. By the time the Cabinet had changed its mind, the enemy landing was a fact and those troops which did enter Thailand were quickly halted by the advancing Japanese. As a result of this initial failure to act decisively the Japanese were able to complete their build-up unopposed and 11th Indian Division found itself deployed for attack rather than defence.

With the exception of isolated incidents, the Japanese Army and Naval Air Forces soon obtained complete mastery of the air and began operating from airfields in Thailand and northern Malaya almost as soon as they were captured. The British belief that Singapore lay beyond the range of Japanese bombers was rudely shattered when enemy planes arrived over the unsuspecting city shortly before dawn on 8 December. By evening, strikes against British airfields had reduced the number of serviceable aircraft in northern Malaya from 110 to 50, and the process continued until the RAF's bomber force, and therefore its capacity to strike back, had been eliminated.

It was, however, at sea that British morale received a blow from which it never recovered. During the evening of 8 December Rear-Admiral Sir Tom Phillips had sailed from Singapore with *Prince of Wales*, *Repulse* and their escorting destroyers, known collectively as Force Z, hoping to intercept and destroy the Japanese invasion convoys despite the fact that he lacked air cover. Force Z was spotted and shadowed by the enemy's reconnaissance aircraft and submarines and at 11:13 on 10 December the Japanese bombers and torpedo aircraft closed in for the kill. By 13:20 the waters of the South China Sea had closed over both capital ships, which had gone down fighting and with heavy loss of life. The cost to the Japanese was a mere three aircraft lost during the attack itself, one wrecked while landing on its return to base, two seriously damaged and twenty-five with lesser damage. As originally conceived, Force Z was to have contained the aircraft carrier *Indomitable*, but she had grounded on a reef the previous month and been unable to make the voyage to the Far East. Even had she been present it is unlikely that the result would have been very different.

Having effectively eliminated British air power and destroyed Force Z during the first days of the campaign, the Japanese could reinforce at will. The 5th Division was, in fact, already advancing south across the frontier, spearheaded by a medium tank company and a lorried infantry battalion. On 11 December, near Jitra, this advance guard found itself confronted by a screen of ten 2-pounder anti-tank guns, lined up across the road. The 2-pounders, although obsolete, was quite capable of defeating any Japanese tank, but the crews of these particular guns were sheltering from the heavy rain under nearby rubber trees. The battery was overrun and the tanks roared on, ploughing their way into the column of a brigade withdrawing to new positions. To the inexperienced Indian infantry all that mattered was that the Japanese tanks were apparently invulnerable and certainly unstoppable. Two battalions were routed and from these only 200 men succeeded in reaching their own lines; the remainder were cut off and forced to surrender.

This engagement set the pattern for what followed. The 11th Indian Division continued to withdraw, attempting to form successive defence lines along the various rivers that flowed down to the west coast. It was soon apparent to Yamashita that he was not opposed by a first-class enemy and that his opponents were mentally road-bound by their mechanical transport. He therefore decided that by maintaining the momentum of his advance he could deny the British time in which to consolidate their new positions, which could be turned or outflanked before they were properly established. To this end his men travelled very light, commandeering thousands of bicycles. Whole units pedalled or pushed their machines along any sort of track that provided reasonable going, or carried them when it did not. The Japanese fed so well from captured British supplies that

Japanese 70mm infantry battalion howitzer. (IWM)

their slender marching rations went untouched. This enabled their supply services to concentrate on the forward delivery of ammunition, using more bicycles, pack animals and impressed labour. Equally important was the reinforced field engineering element possessed by every division, which travelled well forward and was able to repair demolished bridges with startling speed, so permitting tanks and artillery to keep up with the advance.

It was not only the continued momentum of the advance which the British found unsettling, but also the apparent ease with which the Japanese were able to penetrate the areas behind their newly established defences. Yet the answer was very simple. When the Japanese encountered opposition they would leave about one-third of their strength to conduct a holding attack; the remainder would execute a wide loop through the jungle and rejoin the road some miles to the defenders' rear, where they would establish roadblocks. These were normally sited near a bend and were covered by mortar, machine gun and small arms fire. An ambushed vehicle usually formed the basis of the block, to which a local cart or felled tree might be added. Attempts by unarmoured vehicles to drive through were welcomed since they invariably failed and merely added to the jam. If time permitted, several such blocks might be constructed in the space of a few miles. The local British commander, already engaged in holding off a frontal attack which might include tanks, was then faced with the alternatives of either asking friendly troops beyond the block to clear the road, or fighting his own way out. Generally, the latter was the only course open; if he was unable to break the block himself his men might reach safety by filtering through the jungle, but this meant abandoning all his artillery, anti-tank guns, transport and stores. There were occasions, too, when more equipment had to be abandoned on the wrong side of the bridges prematurely blown by engineers unnerved by the pace of the Japanese advance.

While the 5th and 18th Divisions advanced down the peninsula, Yamashita employed part of the Imperial Guards Division to carry out a series of amphibious operations which outflanked the river lines. Throughout the remainder of December the 11th Indian Division was pushed steadily south. It was by now thoroughly demoralized and was so exhausted that it was not functioning properly as a military formation. Percival himself recognized that the men's reactions were now so slow as to be 'subnormal', but instead of relieving them with the fresh 8th Australian Division he chose to leave them in the line. The scene was set for a major disaster.

On 7 January the division, commanded by Brigadier A. Paris since Murray-Lyon had been relieved on 24 December, was holding the line of the Slim River at a point where major road and rail bridges crossed within five miles of each other. Two days previously it had repulsed Japanese probing attacks but little effort seems to have been made subsequently to construct a coherent defence. The field artillery was held in reserve and only one troop of anti-tank guns was allocated to the 12th Brigade, holding Trolak to the north of the river, where the road and railway run close together for a while. The 28th Brigade, responsible for the security of the bridges, had not yet taken up its positions. But perhaps worst of all was the fact that out of the divisional stock of 1,400 anti-tank mines, just 24 had been laid.

The Japanese had originally intended making their usual holding attack while they carried out flank marches through the jungle, using the three battalions of their 42nd Regiment. However, a Major Shimada had requested the honour of leading the assault down the road with his tank company, which contained 17 medium and 3 light tanks. Per-

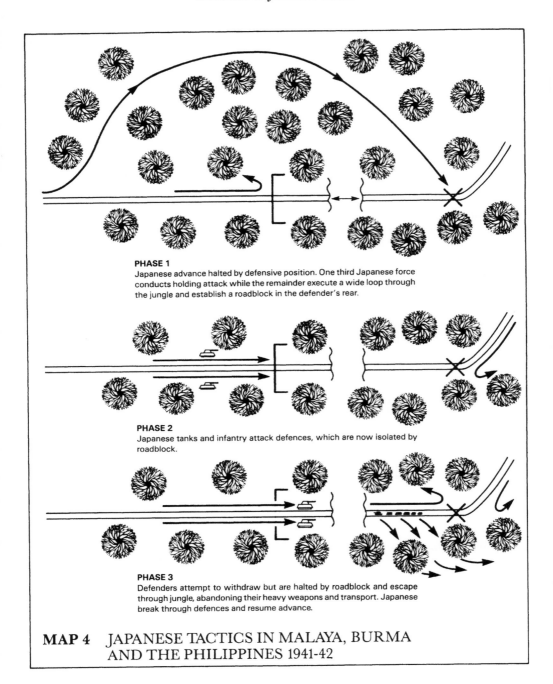

PHASE 1
Japanese advance halted by defensive position. One third Japanese force conducts holding attack while the remainder execute a wide loop through the jungle and establish a roadblock in the defender's rear.

PHASE 2
Japanese tanks and infantry attack defences, which are now isolated by roadblock.

PHASE 3
Defenders attempt to withdraw but are halted by roadblock and escape through jungle, abandoning their heavy weapons and transport. Japanese break through defences and resume advance.

MAP 4   JAPANESE TACTICS IN MALAYA, BURMA AND THE PHILIPPINES 1941-42

mission was granted and a lorried infantry company was also placed under his command. Shimada struck at 03:30, smashing his way through 12th Brigade's battalions, and headed for the road bridge. Here a Gurkha battalion was surprised and virtually destroyed while it was marching towards the bridge. Two artillery batteries parked at the roadside suffered a similar fate. The light anti-aircraft troop guarding the bridge engaged the tanks but was overwhelmed. Crossing the bridge, Shimada pressed on for two miles and at 09:30 ran into an artillery regiment driving north. He inflicted some loss but the gunners, who believed themselves to be 19 miles behind the front, quickly recovered from their initial shock and, unlimbering a 4.5-inch howitzer, blew the leading tank apart at 30 yards range.

Shimada then withdrew to guard the bridge he had so spectacularly captured, being awarded the high honour of a Unit Citation. His losses from all causes amounted to six tanks. That night the remnants of 28th Brigade withdrew south of the river, blowing the railway bridge behind them.

The Slim River débâcle cost the British 4,000 casualties, most of whom were captured, plus 33 medium and field artillery weapons, 15 anti-tank guns, 6 light anti-aircraft guns, 50 tracked carriers and armoured cars, 550 motor vehicles and large quantities of ammunition and food. 12th Brigade had been destroyed and with 28th Brigade no longer fit for action the 11th Indian Division was finished for a while. On the less critical east coast sector the 9th Indian Division was faring somewhat better

Japanese infantry and Type 95 light tanks on the Bataan front, Luzon. (Via Isobe Takuo)

and Percival still hoped that III Corps would be able to make a stand north of Kuala Lumpur, but General Sir Archibald Wavell, who had recently been appointed Supreme Commander of all Allied forces in South-East Asia, otherwise known as ABDA (American, British, Dutch, Australian) Command, did not agree. He was passing through Singapore and personally overruled Percival, insisting that 11th Indian Division should be taken out of the line. Furthermore, since northern Malaya was already lost and central Malaya offered reasonable going for the Japanese armour, a new defence line would be established in Johore and held by the reinforced 8th Australian Division, 9th Indian Division and such elements of 11th Indian Division as were still in a condition to fight. The deteriorating situation had also induced the British Chiefs of Staff to reinforce Malaya with the 17th Indian and 18th British Divisions, the leading brigades of which were to be sent into the line as soon as they reached Singapore; sadly, both formations were untrained and after two months at sea the British were unfit. The veteran 7th Armoured Brigade was also despatched from the Middle East, but was sent to Burma instead.

The Japanese followed up the withdrawal, taking Kuala Lumpur on 11 January and Malacca a week later. The fighting in Johore followed the pattern of that further north, although the Australians staged a number of ambushes which wrote down the enemy's armour and inflicted heavier losses on his infantry. Despite these local successes the tide of battle was now flowing irrevocably against the British. By the end of the month Percival had decided to abandon the Malayan mainland and withdrew his troops across the causeway to Singapore Island.

Yamashita's resources were now stretched to their limits and the few Hurricanes which had reached Singapore were making life so difficult for his motor transport echelons that they were unable to operate in daylight. Nevertheless, he was unimpressed by the quality of his opponents and although they greatly outnumbered his own tired but buoyant troops he judged that they could not withstand a determined assault. Despite the considerable risks involved, therefore, he decided to bring the campaign to a triumphant conclusion by capturing the great prize of Singapore itself.

Wavell thought that the Japanese would land on the island's north-western coast, but Percival felt that the assault would be delivered further east, a belief which Yamashita encouraged with ostentatious troop movements. Percival had not formed a central reserve and intended defeating the landing at the water's edge, despite the fact that little or no work had been done to construct defences along the vulnerable northern coast, the excuse being that it would have seriously affected morale while fighting was still in progress on the mainland. Curiously, no such consideration was given to the wholesale demolitions taking place in the great naval dockyard to prevent its use by the enemy. To the troops who had returned from the mainland, many of whom believed that the Japanese were unbeatable, and to the untrained reinforcements still reaching the port, it seemed that their commanders considered defeat inevitable. Understandably, the possibility of dying in the last battle of a lost campaign was something every soldier wished to avoid. Little fighting spirit remained and the decline in discipline was marked by desertions, disrespect for officers and drunkenness.

As Wavell had predicted, Yamashita had chosen the north-western coast for his assault. Following a heavy bombardment, the assault craft of the Japanese 5th and 18th Divisions crossed the Straits of Johore during the night of 8/9 February, swamping the Australian 22nd Brigade with sheer weight of numbers. It was typical of the malaise afflicting the British conduct of operations that the Australians were forbidden to use their searchlights without the permission of higher authority; this would undoubtedly have been given, had not the all-important field telephone cables been severed by shell-fire. Again, the radios upon which the Australians relied to control their artillery support had, incredibly, been withdrawn for servicing. Next night, the Imperial Guards Division secured a beach-head in the area of 22nd Brigade's eastern neighbour, the Australian 27th Brigade.

During the days that followed the Japanese made steady progress towards Singapore city, particularly after their tanks began coming ashore. By the morning of 15 February they had obtained control of the vital water reservoirs. Percival believed that if the Japanese were forced to storm the city they would massacre the civilian population, just as they had done at Shanghai and Hong Kong, and he asked Yamashita for terms.

The two met that evening in a room at the Ford Motor Company's factory at Bukit Timah. Yamashita was himself on the horns of a dilemma, having only sufficient ammunition in hand for three days' serious fighting, but he bluffed angrily and gave nothing away save a promise to safeguard the lives of troops and civilians, a promise which he kept. It was further agreed that at 21:20 Percival's troops would lay down their arms.

This act set the final seal not only on the greatest military defeat ever sustained by the British Empire, but also on the most decisive jungle campaign in history. British battle casualties amounted to 9,000, but an entire army had been destroyed and 130,000 of its men were now prisoners of war. The Japanese also sustained 9,000 casualties, including 3,000 killed, plus 331 aircraft and perhaps one-third of their tanks. In the wider sphere, British prestige in the Far East was so seriously damaged that Australia and New Zealand turned to the United States for their

defence, and the concept of colonialism received a blow from which it proved impossible to recover. As for Yamashita, he had served his purpose. As soon as the incredulous jubilation had subsided, his enemies quickly removed him to the obscurity of an administrative post in Manchuria; not until July 1944, when Japan's sun was clearly setting, was he recalled to an active command, being made responsible for the defence of the Phillipines.

In Burma the pattern of the fighting was similar, with several significant differences. The Japanese wanted Burma for its oil and rice and because they were determined to sever the Burma Road link with China once and for all. On 15 January 1942 Lower Burma was invaded from Thailand by Lieutenant-General Shojiro Iida's Fifteenth Army, containing the 33rd and 55th Divisions, commanded respectively by Lieutenant-General Shozo Sakurai and Lieutenant-General Yutaka Takeuchi. The British garrison was much smaller than that of Malaya and consisted of the 17th Indian Division (Major-General John Smyth), an untrained formation hastily assembled around the divisional headquarters of those reinforcement brigades which had been despatched to Singapore, and the equally untrained 1st Burma Division (Major-General Bruce Scott), under the overall command of Lieutenant-General T. J. Hutton. Chiang Kai-shek, anxious to preserve the Burma Road, contributed several under-equipped Chinese divisions, although it would be some time before these reached the front, and the RAF's meagre strength was augmented by the famous American Volunteer Group, otherwise known as the Flying Tigers.

British strategy was based on the defence of Rangoon and for this reason the 17th Indian Division was positioned in Lower Burma. After the initial contacts the division fell back to avoid encirclement, hop-

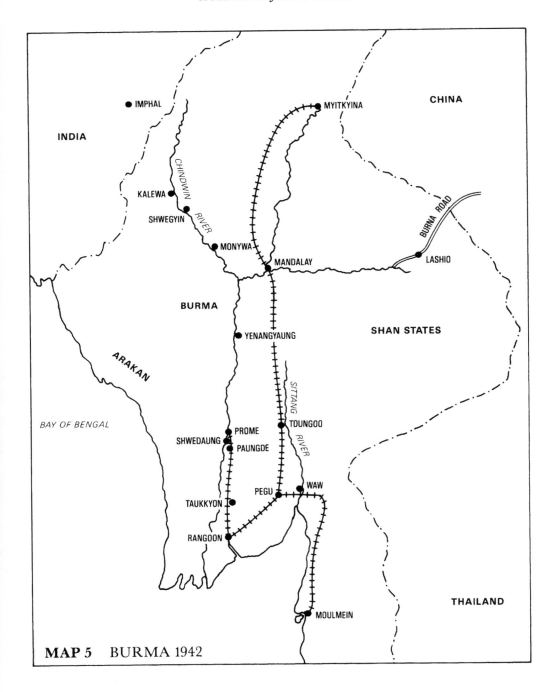

MAP 5 BURMA 1942

ing to hold the line of the Salween River. On 11 February, however, the Japanese 33rd Division secured a bridgehead and the following day the British fell back to the line of the Bilin. Advancing rapidly along jungle trails, the Japanese forded this river and on 18 February compelled a further withdrawal behind the Sittang. Again the Japanese followed up their success and were soon in a position to menace the only road bridge while most of Smyth's road-bound troops were still on the eastern bank. By the early hours of 23 February the defenders of the bridge were under such pressure that at 04:30 the order was given to blow it. Those that could built rafts or swam across, but many more were captured or drowned attempting to reach safety. Most of the divisional artillery and transport were lost, as

was a high proportion of small arms. Only 4,000 men reached the rally-points, and for these only 1,420 rifles and 56 light machine guns were available.

17th Indian Division fell back on Pegu, north-east of Rangoon, to reform. In the aftermath of the Sittang disaster Smyth was replaced as divisional commander by Brigadier David Cowan and, at the higher level, Hutton was relieved by Lieutenant-General the Hon Sir Harold Alexander. Fortunately, the Japanese were extremely tired and had outrun their supplies, and a pause ensued during which the British were able to bring forward Brigadier J. Anstice's 7th Armoured Brigade, which had landed at Rangoon on 21 February. This was the most experienced formation in the British Army, having served in the Western Desert from the start

Stuart tanks of the 2nd Royal Tank Regiment in a harbour area during the retreat form Burma. Had it not been for the presence of 7th Armoured Brigade, which had originally been destined for Malaya, the British army in Burma would probably have been destroyed. (2nd Royal Tank Regiment)

of the North African campaign, and it had been bound for Singapore when news of Percival's surrender led to its being diverted to Burma. Its two regiments, 7th Queen's Own Hussars and 2nd Royal Tank Regiment, each possessed a full complement of M3 Stuart light tanks, which were extremely reliable and armed with a 37 mm gun capable of defeating their opponents. The brigade's first task was to provide cover for 17th Indian Division while it re-formed, the first squadron moving into position at Waw, east of Pegu, on 23 February.

The Japanese renewed their offensive on 2 March. Their infantry suffered severely from the Stuarts' close-quarter fire and when their own tanks arrived on the scene it was only too obvious that their crews knew nothing about fighting a tank battle. Four

Type 95s were destroyed in two engagements, and a fifth was hastily abandoned. After this, the Japanese tankmen avoided direct contact with the Stuarts. Unfortunately, the British armour could not be everywhere and large numbers of the enemy flowed round the battle, their objective clearly being Rangoon. On 6 and 7 March 7th Armoured and 48th Brigades, which formed the rearguard of 17th Indian Division, were forced to fight their way back through the blazing town of Pegu and roadblocks beyond.

Wavell had been adamant that Rangoon should be held, and Alexander was most reluctant to relinquish the port, since its loss would mean the virtual isolation of his army, land communications with India being limited to a few all-weather tracks through

Left *The Japanese 215th Infantry Regiment crosses the Thai–Burma border, January 1942.* (215th Infantry Regiment War Veterans Association)

Above right *Banzai! Members of 215th Infantry Regiment celebrate their capture of Rangoon, 10 March 1942.* (215 IRWVA)

the jungle-clad hills beyond the Chindwin. Nevertheless, it was now apparent that holding the city would simply present the Japanese with a triumph similar to Singapore and on 6 March Alexander decided to abandon it and withdraw to the north. This operation in itself was difficult as it involved backloading all the army's stores and establishing dumps at suitable locations along the way.

During 7 March the army commenced its withdrawal. At Taukkyon, some 24 miles north of Rangoon, the column was brought to a standstill by a stubbornly defended roadblock. This had been established by a Major Takonubu of the III/214th Regiment to protect the flank of the 33rd Division as it crossed the road to assault Rangoon from the west. Although he was unaware of the fact, Takonubu had trapped most of the 17th Indian Division, 7th Armoured Brigade and, worst of all, Alexander and his headquarters staff; in addition, thousands of refugees and countless civilian vehicles were adding to one of the worst traffic jams of the war. Attacks against the block from north

and south all failed. A major all-arms attack in brigade strength was planned for dawn the following day, but at first light reconnaissance revealed the inexplicable fact that the Japanese had gone. Sakurai had never imagined that Alexander would abandon Rangoon and was thus unaware of the prize that he held within his grasp; having successfully completed his mission, Takonubu had conformed to the Japanese tradition of rigid obedience to orders and marched off as soon as 33rd Division's rearguard crossed the road.

Once Rangoon had been secured, Iida received substantial reinforcements. These included Mutaguchi's 18th Division, fresh from its triumphs in Malaya, and Lieutenant-General Sukezo Matsuyama's 56th Division, which had been held in reserve during the Malayan campaign, plus two tank regiments, more artillery and more aircraft. To protect their seaborne communications the Japanese sent their carrier fleet into the Indian Ocean during the first week of April, raiding the naval bases at Colombo and Trincomalee in Ceylon, and sinking the carrier *Hermes*, the cruisers *Cornwall* and *Dorsetshire* and several lesser warships at sea, so forcing the Royal Navy to abandon the Bay of Bengal.

Meanwhile, the British forces had been concentrated in the area of Prome in the Irrawaddy valley and redesignated Burcorps, command of which was assumed by Lieutenant-General William Slim on 19 March. Further east, the Chinese Fifth and Sixth Armies, each smaller than a British corps, assumed responsibility for the Sittang valley sector and the Shan States under the command of Lieutenant-General Joseph Stilwell, US Army. The problem was that neither Slim nor Stilwell could maintain their positions if the other was forced to withdraw, and as the Chinese were seriously underequipped the prospect of Burcorps being able to hold its ground was remote, although this was not yet fully accepted by its aggressive new commander. Iida, for his part, regarded the Chinese as being the more dangerous opponents and eventually concentrated his divisions against them, leaving only Sakurai's 33rd Division to follow up the British.

On 22 March the remnants of the Allied air force, which had fought hard and inflicted heavy losses, were eliminated in a series of raids on Magwe airfield; after this, the Japanese enjoyed complete air superiority for the rest of the campaign. At Toungoo the Chinese came under extreme pressure and to relieve this Alexander ordered Burcorps to mount a counter-attack on 28 March. This involved part of the 17th Indian Division and 7th Hussars and was directed at Paungde. Unfortunately, Sakurai's troops were also advancing and these cut in behind the attackers, establishing a formidable roadblock at Shwedaung. With extreme difficulty the counter-attack force fought its way out at a cost of 300

One of the Essex Yeomanry's 25-pounder guns lost during the abortive counter-attack at Shwedaung, March 1942. (215 IRWVA)

casualties, almost all of its transport, two guns and ten tanks, some of the last being repaired and taken into service by the Japanese. As well as being entirely abortive, the demonstration had no effect on the situation at Toungoo, where the Chinese were overwhelmed after a fierce struggle. Burcorps, therefore, had no alternative other than to conform to the withdrawal of its allies.

While carrying out extensive rearguard patrols, 7th Hussars had several brushes with Thai troops, Thailand having thrown in her lot with Japan, and on a number of occasions the regiment's tanks were attacked at close range with frangible glass grenades containing liquid hydrogen cyanide. The idea was that when the glass broke the liquid would vaporise and be drawn into the vehicle, incapacitating the crew. Its success was very limited and it does not seem to have been pursued beyond the period of Japanese expansion.

On 15 April the vast oilfield at Yenaungyaung was destroyed by the British to prevent its use by the enemy, but the following evening the fast-moving Japanese ambushed Burcorps' rearguard north of the town, cutting its column into several sections. Most of 1st Burma Division, with A Squadron 2 RTR under command, were trapped south of the Pin Chaung (chaung is the Burmese word for watercourse, inland or tidal), the banks of which were held by the Japanese in strength. North of the chaung was the remainder of 2 RTR, Headquarters 7th Armoured Brigade and supporting troops, and although these elements were able to restore the situation locally they were unable to regain contact with 1st Burma Division. Alexander at once arranged with Stilwell that the newly arrived Chinese 38th Division, commanded by Lieutenant-General Sun Li-jen, who had received his education

Wrecked 7th Hussar Stuart at Shwedaung. The vehicle is almost certainly that commanded by Lieutenant Kildare Patteson, who was captured by the Japanese and tied to a wrecked ambulance. The position was then shelled by the British artillery, but Patteson managed to free himself and escape. (215 IRWVA)

Another view of the roadblock at Shwedaung, showing burnt-out scout cars and an abandoned 7th Hussar Stuart. (215 IRWVA)

at an American military academy, should be allocated to Slim for the relief operation. The division lacked artillery, transport and indeed heavy equipment of every kind, but its morale was high and it was decided that it would attack across the Pin Chaung at 06:30 on 18 April, supported by 2 RTR. The Chinese were late starting and, to the amusement of the British tank crews, had put back their watches one hour to save face. The attack made some progress and was resumed the following day, capturing a ford across the chaung and reaching the outskirts of Yenaungyaung. Meanwhile, a friendly Burmese had informed Major-General Scott, commanding 1st Burma Division, of an obscure jungle route round the enemy's eastern flank and when A Squadron 2 RTR

verified that this was clear the division, which had suffered severely during the three-day battle and at one time had almost surrendered for lack of water, came out on foot, having destroyed its guns and transport. The wounded, too, had to be left behind and the Japanese, balked of their prey, vindictively butchered them in their ambulances.

During the next few days Iida, having concentrated his 18th, 55th and 56th Divisions, launched a major offensive against the Chinese Fifth and Sixth Armies, routing them and cutting the Burma Road at Lashio on 29 April. As the remnants of the Chinese armies disintegrated and fled northwards towards their own frontier, Alexander decided that Burcorps and Sun's 38th Divi-

46

215th Infantry Regiment resumes its march from Shwedaung to Prome. Note animal transport. (215 IRWVA)

sion would withdraw to India, a decision which was inevitable anyway, given that the supplies backloaded from Rangoon would not last forever. On 30 April Slim's rearguard passed through Mandalay, which had been devastated by air attack, crossed the Irrawaddy and blew the Ava bridge. The corps then retired towards the Chindwin, harried by Sakurai, who had begun transferring troops to the west bank of the Irrawaddy on 25 April, the capture by these of Monywa on 31 May forcing the majority of Slim's command to converge on the alternative crossing site of Shwegyin, further north. Here a small fleet of river steamers had been assembled to transport the troops upstream to Kalewa and safety. Before the evacuation had been completed the Japanese

occupied the bluffs overlooking Shwegyin, whence counter-attacks failed to dislodge them. At about 17:00 on 10 May the remaining guns fired off their ammunition in a whirlwind bombardment, the intensity of which temporarily stunned the enemy. The weapons were then wrecked, as were 7th Armoured Brigade's tanks, the latter so thoroughly that when a former officer of the brigade passed the spot over two years later not one had been moved. Having completed the sorry work of destruction the men marched up-river to Kaing and there crossed to Kalewa. The arrival of the monsoon inhibited any pursuit the Japanese may have intended and Burcorps retired unmolested to Imphal.

So ended the longest retreat ever undertaken by a British army. Burcorps had withdrawn 1,000 miles and, while defeated, had retained its cohesion to the end. It had saved 28 guns and some of its transport. One tank had also been brought across the Chindwin, towed on a raft behind a river steamer; when, in due course, the British re-conquered Burma, it would take part in the final advance on Rangoon. The corps' casualties amounted to 4,000 killed and wounded and over 9,000 missing, but included in the latter were many Burmese soldiers who had simply gone home. That Burcorps had survived at all was due largely to the presence of 7th Armoured Brigade, which had not only inflicted heavy casualties but also denied the enemy armour the sort of runaway successes it had enjoyed in Malaya. Without the brigade's Stuarts it is probable that the corps would have shared the fate of the tankless Chinese armies, whose losses undoubtedly exceeded those of the British. The Japanese had employed more troops in Burma than they had in Malaya, but there could be no denying the completeness of their victory; Sakurai in particular had fought a brilliant campaign, granting his opponents little respite. In total, Japanese losses came to 4,597 killed and wounded.

In the Philippines the deteriorating relations between Japan and the United States had led to the Filipino armed services being placed under American command as early as July 1941. By December of that year the Commander-in-Chief, General Douglas MacArthur, could muster 22,400 American regulars, 3,000 Philippine Constabulary and a 107,000-strong Philippine Army which had recently been expanded to the point at which it had become a *levée-en-masse*, inadequately organised, trained, armed or psychologically prepared for war. In addition, MacArthur had available two M3 Stuart light tank battalions, which were formed into a Provisional Tank Group, and one M3 75 mm tank destroyer battalion. Most of his

strength was deployed on the principal island of Luzon, with Major-General Jonathan M. Wainwright's North Luzon Force of four divisions north of Manila where it covered the likely invasion area of Lingayen Gulf, South Luzon Force of two divisions under Brigadier-General George M. Parker south of Manila, and one division in reserve; elsewhere, three divisions were based in the southern islands. Major-General Lewis H. Brereton's Far East Air Force, which included 35 B-17 heavy bombers, was located at the Clark Field airbase north of Manila, but most of the small US Asiatic Fleet had been withdrawn to Java. In the event of invasion MacArthur's strategy bore some similarity to Percival's in that he intended fighting a holding action and, if necessary, withdrawing into the Bataan Peninsula, forming the northern arm of Manila Bay, where he would hold out until the Pacific Fleet arrived with reinforcements.

The Japanese intelligence service was fully aware of MacArthur's dispositions and had formed so low an opinion of his troops that Lieutenant-General Masaharu Homma's Fourteenth Army, detailed for the invasion, consisted only of the 16th and 48th Divisions, two tank regiments, a medium artillery group, three engineer regiments and five anti-aircraft battalions.

The battle for the Philippines began at 12:15 on 8 December 1941 with a heavy Japanese air strike at the Clark Field—Iba airbase complex. This destroyed half the B-17 bomber force which MacArthur had intended launching against Japanese bases in Formosa, plus 56 fighters and 26 other aircraft. Further air raids wrecked the facilities of the naval base at Cavite. Between 10 and 22 December small landings were made at Vigan and Aparri in northern Luzon, at Legaspi in the south of the island, at Davao on Mindanao, and on Jolo Island, the object being to seize airfields to which supporting air squadrons could be flown from Formosa;

MAP 6 LUZON, PHILIPPINE ISLANDS
BATAAN PENINSULA

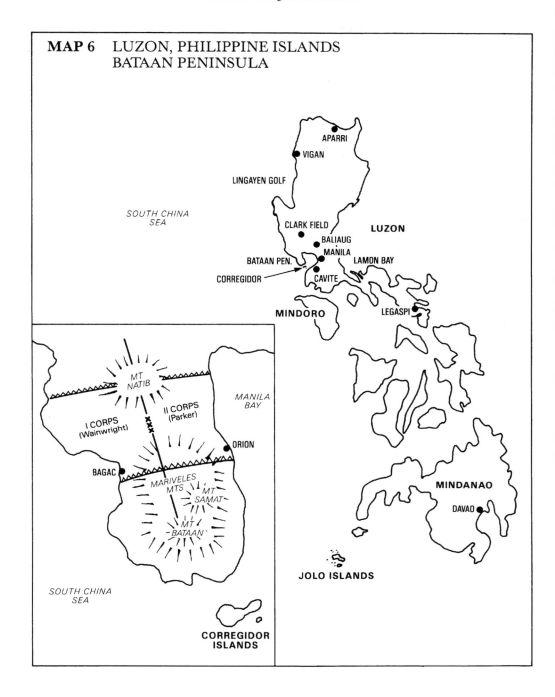

all succeeded against varying degrees of local opposition. The main landing, as MacArthur had predicted, took place in Lingayen Gulf, where the 48th Division came ashore on 22 December. Two days later the major part of the 16th Division landed at Lamon Bay on the east coast. Homma's strategy now became clear. His two divisions were to converge on the central plain, where MacArthur's army would be destroyed in the area of Manila. MacArthur, however, was aware of the danger and had already decided to withdraw into the Bataan Peninsula, declaring Manila an open city.

The early stages of the withdrawal were conducted in relatively open, developed country. The combination of frontal and flank attacks, coupled with the ability of the Japanese to maintain the momentum of their offensive, was as unsettling to the defenders of the Philippines as it was to those of Malaya. Bridges were blown prematurely by nervous engineers, and because of this the Provisional Tank Group lost about one-third of its tanks before the withdrawal had been completed. More were lost as a result of inexpert handling by untrained crews, but on 31 December the Group had some measure of revenge when it destroyed eight Type 95s of the Japanese 4th Tank Regiment at Baliuag.

Once his troops had been concentrated on the mountainous, jungle-clad peninsula, MacArthur could take stock. The last of his B-17s had flown out to Australia on 17 December and the enemy possessed complete air superiority. The only American naval vessels remaining were a handful of minesweepers, gunboats and motor torpedo boats, and the elimination of the Pacific Fleet at Pearl Harbor meant that the prospects of relief were remote indeed. Sufficient food had been stockpiled on Bataan to feed 43,000 men for six months, but some 80,000 troops were present, plus 26,000 civilian refugees from the Manila area, and this

meant that rations would have to be cut by half. Nor were the medical facilities adequate to cope with the growing number of malaria cases. On the other hand, his flanks now rested securely on the sea and the Filipino element of the army had not only settled down well after the first shock of battle, but also shown itself to be capable of offering the most determined resistance. An outer defence line was established on either side of the precipitous slopes of Mount Natib, while eight miles to the south a shorter inner line was constructed between Orion and Bagac. The western sector of these defences was held by Wainwright's I Corps (1st, 11th and 91st Divisions), and the eastern sector by Parker's II Corps (21st, 41st, 51st and part of 31st Divisions).

Having marched in triumph through the deserted streets of Manila, Homma returned to the attack. A major offensive was launched on 9 January 1942 and despite bitter fighting in which the well-handled American artillery inflicted heavy losses, by the end of the month both corps had been forced back to the inner defence line. An attempt to land two battalions behind American lines on the south-west coast of the peninsula was foiled by local counter-attacks, motor torpedo boats and gunfire from the island fortress of Corregidor, although it took three weeks to destroy the last elements of these.

For the moment, however, Homma had shot his bolt and the arrogance of the Imperial Staff in allocating just two divisions for the conquest of Luzon, and allowing a mere 50 days for its completion, stood starkly revealed. The Fourteenth Army had sustained 7,000 battle casualties, including 2,700 killed, but even worse was the fact that over 10,000 of its men had gone down with tropical diseases. If MacArthur's troops had had the capacity they could have rolled over the Japanese, but they were in little better state and the time was approaching when their rations would again be cut by half. Even so, Homma was forced to suspend operations until the fresh 4th Division could be shipped from Shanghai. During the lull MacArthur, reluctantly obeying a direct order from President Roosevelt, left the Philippines for Australia on 11 March. His responsibilities were assumed by Wainwright, and Major-General Edward P. King was appointed commander on Bataan.

By the end of March the number of American and Filipino sick had risen to 24,000. On 3 April Homma's reinforced army commenced its final offensive, concentrating the fire of 150 guns, supplemented by heavy bombing attacks, against a comparatively narrow sector of II Corps' front north of Mount Samat. This resulted in limited gains and was repeated the following day, breaking the morale of 41st Division, some units of which fled to the rear. The Japanese swarmed through the gap, beating off weak counter-attacks, and as units on either flank withdrew the whole line began to disintegrate. King recognized that his men, listless from hunger and sickness, demoralised and disorganised, were incapable of further effort and surrendered on 9 April. Some 2,000 fugitives managed to reach Corregidor but the remaining 78,000 were marched 65 miles under a broiling sun to their prison camp at San Fernando. During this, the infamous Death March, little food or water was supplied, the hundreds who fell by the wayside being bayonetted or clubbed to death; hundreds more never recovered from the ordeal.

Corregidor continued to resist. On 5 May, however, the Japanese overran most of the shell- and bomb-blasted island and Wainwright surrendered the following day, instructing his troops in the southern Philippines to conform. If the Japanese believed that that was the end of the matter, they were sadly mistaken. Everywhere, American and Filippino soldiers escaped into the mountainous interior of the islands to become the nucleus of a popular and well-organised resistance movement which was supplied with arms by submarine and which was to tie down badly needed Japanese troops in a guerrilla war for the next three years.

Despite the immensity of the area involved, the Japanese Sixteenth Army's conquest of the Dutch East Indies, and of the British possessions of Sarawak, Brunei and North Borneo, was a much simpler matter. The Dutch, isolated from their German-occupied homeland, were poorly equipped and the majority of their 85,000 troops consisted of colonial units whose loyalty was questionable. Even after the formation of ABDA Command and the provision of limited British, Australian and American reinforcements, it was impossible to defend so many islands with the resources available, however thinly they were spread. For the Japanese it was merely necessary to possess overwhelming strength in the relevant areas, and this they were able to achieve by means of sea power and air superiority, the latter enhanced by the acquisition of airfields in Malaya and the southern Philippines.

These factors enabled them to effect landings more or less at will throughout the entire area, their primary objects being to secure airfields with which to support their further advance and the capture of oil installations before they could be destroyed, and to these ends paratroops were sometimes dropped shortly before the arrival of a seaborne invasion force. Local garrisons might offer determined resistance for a while, but their isolation ensured that they were soon crushed. Starting with landings at Miri and Kuching in Sarawak in December, the Japanese moved steadily southwards from point to point throughout January and February, capturing Sandakan in British North Borneo, Tarakan, Balikpapan, Bandjermasin, Pemnagkat and Pontiakat in Borneo, Manado, Kendari and Makassar in the

Celebes, Dili and Kupang on Timor, Ambon, Banka and Bali Islands, and Palembang in Sumatra. On 27 February an Allied naval squadron attempted to intercept an invasion convoy bound for Java but was defeated in the Battle of the Java Sea. The Japanese landed at several points on the island's north coast and on 8 March the Dutch Governor-General surrendered.

The Rising Sun now stood at its zenith. Japan had acquired all the natural resources she sought and her conquests stretched from the frontier of India to the Central Pacific. However, the more thoughtful of her senior officers, notably Admiral Isoroku Yamamoto, the Commander-in-Chief of the Imperial Navy, were uneasy. The Western Allies showed no signs that they accepted the verdict of the short victorious war, nor that they sought a negotiated settlement. Japan's armament industry could not hope to equal the enormous output of that of the United States and the consequences of a prolonged war would be disastrous for the Empire. One short month after Wainwright had surrendered Corregidor, four aircraft carriers, the pride of the Imperial Navy, were sent to the bottom of the Pacific at the Battle of Midway, and with them went their superbly trained aircrews. The Red Sun of Nippon was already in decline.

CHAPTER 3

LEARNING THE BUSINESS

The sheer immensity of the area conquered by Japan between December 1941 and May 1942 in itself posed enormous problems for the Imperial Staff, which was forced to conclude that, if the vast perimeter was to be adequately defended, further conquests of strategically important territories were necessary. These included New Guinea/ Papua, whence northern Australia could be menaced, and the Solomon Islands, lying on the flank of American communications with Australia and New Zealand, both of which were mountainous and covered with the most venomous jungles to be found on earth. Both also contained indigenous populations which the Japanese contemptuously dismissed but which remained staunchly loyal to the Allies and made an invaluable contribution to their ultimate victory.

Strategic surprise was, of course, no longer possible, and tactical surprise would be difficult to obtain because of a remarkable organisation established by the Australian government under Commander Eric Feldt, RAN. This was known as the Coastwatcher Service and drew its recruits from former planters, traders, colonial officials and officers who volunteered to remain behind when the Japanese invaded their territories. Assisted by loyal natives, the coastwatchers remained in hiding in the jungle, reporting Japanese naval, air and troop movements by radio as they occurred, the advance warning provided often being criti-

cal to the outcome of an engagement. If they were caught, they could expect to be savagely tortured before they were killed; if they remained at liberty, there was always the prospect of a lonely death from tropical disease. Once they were aware of their existence, the Japanese spared no pains to track them down, using radio-location sets to pinpoint their whereabouts. The coastwatchers were therefore compelled to change their positions regularly and in the process became expert bushmen, taught by their native guides. They learned, for example, not to use the principal paths since this avoided leaving their own tracks yet left those of the enemy plain to read and interpret. They learned to walk on stones and roots whenever possible, leaving no mark; on softer surfaces the coastwatcher went first, the distinctive imprint of his boots being obliterated by the bare feet of the guides. They learned, too, from men who had barely emerged from the Stone Age, how to use the resources of the jungle itself to provide warning devices and booby traps, and in due course much of this lore was absorbed by the Allied armies.

The prime objective of the Japanese was Port Moresby on the south coast of New Guinea. This was to have been taken by amphibious landing but on 7 May the carrier force covering the invasion fleet was intercepted by an American carrier squadron. The result was the drawn Battle of the

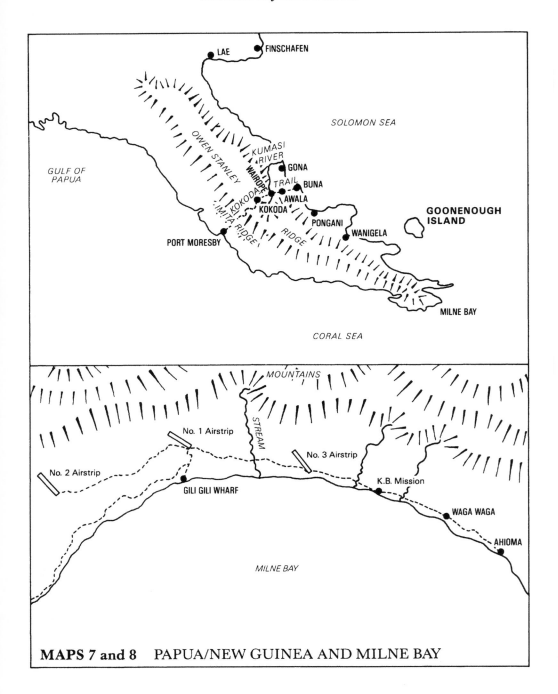

MAPS 7 and 8   PAPUA/NEW GUINEA AND MILNE BAY

Coral Sea, in which the Japanese lost the small carrier *Shoho* and the greater number of aircraft, plus the fleet carrier *Shokaku* severely damaged, while one American carrier, the *Lexington*, a destroyer and an oiler were sunk and the other carrier, the *Yorktown*, sustained some damage. The operation was cancelled and instead it was decided that Port Moresby would be captured by converging thrusts from Milne Bay, at the eastern tip of New Guinea, and from Buna on the north coast across the Owen Stanley Range.

Milne Bay was held by two Australian brigades, the veteran 18th, which had seen active service in the Middle East, and the 7th, consisting of militiamen, under the overall command of Major-General Cyril Clowes, the strategic importance of the area being enhanced by several airstrips on which an American engineer regiment was working. On 24 August a coastwatcher reported seven landing barges off the northern coast, moving in the direction of Milne Bay. These were attacked by the airstrips' Kittyhawks

the following day, driven ashore on Goodenough Island and set ablaze, leaving the 350 infantrymen aboard stranded. The main invasion fleet was spotted and attacked but succeeded in reaching Milne Bay, where it landed a regimental-sized group and several light tanks at Ahioma on the eastern shore of the bay during the night of 25/26 August.

The Japanese began advancing eastwards immediately but found their movement restricted to the narrow strip of land between the sea and mountains clad in dense jungle. They attacked only at night, generally behind tanks which used their lights to illuminate the Australian positions and subjected the ground ahead to a continuous hail of machine gun fire. The militia brigade resisted fiercely but was forced to give ground, being unable to halt the tanks because its anti-tank guns could not be brought forward through the mud. A Boys anti-tank rifle was found and this knocked out two tanks at point-blank range; more

Japanese Type 95 light tank captured during the fighting at Milne Bay. According to Australian accounts, the Japanese tanks used their headlights to dazzle the defenders during night attacks, yet headlights were not fitted to this or other tanks abandoned by the Japanese. On the other hand, the main armament has been stripped out and the turret may have fitted with a light projector. (IWM)

tanks bogged down and the sting was drawn from the enemy's attack.

By 31 August, the Japanese, having been reinforced by 800 Naval Infantry, had reached the edge of No 3 airstrip, which they attempted to capture in a series of unscientific *banzai* charges. These were shot flat and Clowes committed his regular brigade to a counter-attack. Sustained progress was maintained during the next few days, pushing the landing force back towards its base area. During the night of 5/6 September Japanese ships entered the bay and small craft could be heard plying between them and the shore. Dawn revealed that the enemy had gone.

The fighting at Milne Bay, while small in scale, gave the Allies their first clear-cut land victory over the Japanese. Australian losses amounted to 161 killed and missing and about the same number wounded; over 600 Japanese were killed. Possession of local air superiority had been of immense assistance to the Australians and demoralised their opponents, whose positions and base area had been strafed continuously during the hours of daylight. The Australian artillery had also been well handled and the untried militia infantry had fought like veterans. As for the Japanese, they had been confronted by a defence in depth against which they had been unable to develop their usual flanking movements and, despite the possession of tanks and the use of night attacks, their frontal assaults had failed.

Meanwhile, the second and major thrust against Port Moresby had already begun. On 21 July Major-General Tomitoro Horii's South Seas Detachment, consisting of veterans from China, Malaya and the Philippines, began landing at Buna on the north coast. Horii intended advancing south across the Owen Stanley Range using the 100-mile Kokoda Trail, of which he knew so little that one of his officers was detailed to put it into a condition capable of supporting motor transport. The route was, in fact, nothing more than a narrow native trail which wound its way over the mountains, climbing a succession of ridges and descending into intervening valleys, crossing swift-flowing rivers and following crests until the summit was reached, whence it began to drop steadily towards the opposite coast. The ascent from the south was, in places, so steep that it was necessary to crawl and on the worst section, known as the Golden Staircase, the Australians constructed steps from pegged logs. Together, incessant rain and heavy usage quickly turned the trail into a quagmire. Even for fit men the physical effort required in such going was exhausting, but for the many on both sides who were suffering the effects of tropical illness the trail became a living nightmare. On the Kokoda Trail, named after the village half-way along it which contained an administrative centre and airstrip, distances were measured not in miles but in hours and days.

Horii, who had assembled some 13,000 troops, began moving inland shortly after he had landed. He had brought with him over 400 pack horses, which were of limited value, and 1,200 native porters from Rabaul. Attempts to recruit more porters locally met with little success. The first clash occurred near Awala, held by advance units of the Australian 30th Brigade, on 23 July. After inflicting some losses the Australians withdrew behind the Kumusi River, destroying the suspension bridge at Wairopi (Wire Rope). The Japanese engineers bridged the river the following day and the advance continued. The Australian strategy was to inflict maximum casualties and delay, and since the front was of necessity narrow, it was possible to achieve both. Thus, while Horii continued to advance, he did so slowly and at a heavier cost than had been anticipated. As the Australians fell back they were reinforced by the 21st Brigade at the end of August, and then by the 25th Brigade, clad

A 'short' 25-pounder gun-howitzer, developed by the Australian Army specifically to meet the appalling conditions in New Guinea, earns the approval of American artillerymen. (USAMHI)

for the first time in jungle greens.

On 17 September the Japanese reached Imita Ridge and were within 30 miles of Port Moresby, having sustained the loss of 1,000 killed and 1,500 wounded, over three times that of the Australians. Their logistic system had been unable to cope with the demands of the long advance, they were suffering from disease and, since the Australian food dumps had been deliberately backloaded, they were on the verge of starvation and utterly exhausted. They were now confronted by a formidable defensive position which included two 25-pounders, specially shortened for jungle warfare and man-handled up the track with infinite effort. Further attacks would almost certainly have failed but in any event Imperial General Headquarters in Tokyo, depressed by events on Guadalcanal, the failure of the Milne Bay

operation, and worried by the prospect of an Allied landing at Buna, sent Horii a signal on the night of 24 September ordering him to withdraw from the Owen Stanley Range and concentrate around his base.

His retreat along the trail was followed up by 25th and 16th Brigades, which fought their way through several rearguard positions. Here and there unmarked but fresh corpses marked deaths from starvation, and examination of faeces revealed that many of the enemy had been reduced to eating grass and wood; elsewhere, ants had reduced the dead from the earlier fighting to gleaming skeletons. In a dismal valley bottom lay the remains of Horii's white charger; the general himself was drowned attempting to cross the Kumusi River on a raft.

The Australian advance, like the withdrawal which had preceded it, was supported

Left *Australian infantry and native stretcher bearers on the Kokoda Trail.* (IWM)

Right *American infantry cautiously approach a Japanese bunker at Buna. Some idea of the bunker's construction can be made out at the entrance.* (USAMHI)

by thousands of native porters who not only delivered supplies forward but also assisted in the evacuation of the wounded. There were never enough of these hardy, loyal and courageous men who, like Gunga Din, wore 'nothin' much before, an' rather less than 'arf o' that be'ind,' and without them nothing could have been accomplished. Latterly their efforts were supplemented by parachute supply drops, which local air superiority had permitted the Allies to develop.

The fighting at Milne Bay and along the Kokoda Trail had provoked acrimonious discussions among the more senior Allied commanders. From the comfortable surroundings of his Brisbane headquarters, MacArthur, now Commander-in-Chief South-West Pacific Area, had issued a number of statements criticising the conduct of operations in New Guinea, despite the lack of any personal knowledge of the conditions in which the fighting was taking place. In particular, he seemed to regard the low Australian casualties as being evidence of poor fighting potential. Naturally, this was resented by Lieutenant-General Sydney

Rowell, commanding New Guinea Force, and the relationship between the two deteriorated to the point at which General Sir Thomas Blamey, the commander of Australia's land forces, was compelled to replace the latter at the end of September with Lieutenant-General Edmond Herring, although the campaign continued along the lines already established.

Many Australians, who had yet to benefit from MacArthur's strategic insight, felt that for a defeated general he had rather too much to say. The Commander-in-Chief, however, was about to commit American troops to the fray in the form of Lieutenant-General Robert L. Eichelberger's US I Corps, the leading formation of which, 32nd Division under Major-General Edwin F. Harding, had already reached New Guinea. While the 7th Australian Division, commanded by Major-General G. A. Vasey, closed in on Buna along the Kokoda Trail, some of Harding's men marched across the island by a parallel track to the east while the rest were flown to airstrips at Wanigela and Pongani on the north coast. The Allies then

converged on the Buna perimeter; the Australians on the left, the Americans on the right.

Following Horii's death the Japanese forces in New Guinea were designated the Eighteenth Army and placed under the command of Lieutenant-General Hatazo Adachi, who established his headquarters at Rabaul. Reinforcements were rushed to Buna, bringing the number of troops present to 12,000. Long before the Australians had debouched from the Kokoda Trail, work had begun on the construction of extensive field fortifications around the perimeter. These took the form of bunkers sited in depth and with interlocking arcs of fire. The bunkers were reinforced with earth-filled oil drums and roofed with several layers of palm logs and earth so that only a direct hit with a heavy-calibre round would destroy them. They were then expertly camouflaged and the fire slits masked with transplanted shrubs; numerous instances are recorded of Allied troops standing on top of such bunkers without realizing that they were there. The bunkers might be connected by crawl tunnels and, in hilly country, their entrances could be sited on the reverse slopes. Most dangerous of all was the fact that, rather than try and escape if overrun, the Japanese were prepared to die in their bunkers.

By the middle of November the Allies were ready to commence their assault. It failed, and the inexperienced Americans performed so badly that a furious MacArthur sent Eichelberger forward to assume personal control of the 32nd Division. Writing ten years after the event, Eichelberger commented that, in his opinion, the men were more frightened of the jungle than of the Japanese. 'There is nothing pleasant about sinking into a foul-smelling bog up to your knees. There is nothing pleasant about lying in a slit-trench half-submerged, while tropical rain turns it into a river. Jungle noises were strange to the Americans—and in the hot moist darkness the rustling of small animals was easily interpreted as the stealthy approach of the enemy.'

By degrees and leading from the front, Eichelberger succeeded in restoring his men's confidence, but a number of technical aspects remained to be resolved. For example, neither bombs nor shells made much impression on the Japanese bunkers and the artillery's forward observation officers were unable to function efficiently because their vision was obscured by head-high kunai grass. Shooting improved considerably when Wirraway training aircraft were used to spot for the guns and the issue of delayed-action fuses increased their effectiveness against bunkers: whereas previously shells had burst harmlessly on impact with the top layer of palm logs, they now ploughed their way through the roof before exploding.

On 9 December the Australians captured Gona, lying on the northern flank of the enemy perimeter, but it was clear that Buna would only be taken by hard, attritional fighting. Reinforcements arrived from Milne Bay and with them came Stuart light tanks of the Australian 2/6th Armoured Regiment. Few areas offered less promising going for armour. Snipers forced tank commanders to operate closed-down, with the result that their vision was sharply curtailed. Some tanks became bogged in the treacherous semi-swamp, and others bellied immovably on invisible palm stumps; again, the short range at which the fighting took place enabled the unseen enemy guns to penetrate the thin armour, so that casualties in vehicles and crews were comparatively high. Nevertheless, the presence of the Stuarts was critical, for they were able to pump 37 mm and machine gun fire through bunker fire-slits,

Stuarts of the Australian 2/6th Armoured Regiment near Buna. Despite being employed over the worst going imaginable, the tanks made a valuable contribution to the fighting. The leading vehicle has bogged down and is being towed out; an additional girdle of armour has been welded around the turret ring to prevent jamming by shell splinters and small-arms fire. (US Army)

unlocking areas which might have held up the infantry for days. In this way the Allies gnawed their way through the defences until Buna fell on 2 January. The last pocket of resistance, at Sananda some distance to the north, was overwhelmed on 22 January.

Some 7,000 Japanese died at Buna. About 1,200 sick and wounded were evacuated by sea and 1,000 fugitives managed to reach Lae. The Allies incurred 6,410 casualties, plus a high proportion of men incapacitated by sickness. Together, the victories at Milne Bay, on the Kokoda Trail and at Buna removed the threat to Australia, while for the Japanese New Guinea became a lost cause which would claim 100,000 lives. Throughout the next 18 months MacArthur relentlessly pursued his advantage with a series of landings along the north coast of New Guinea, isolating the battered and disorganized Japanese Eighteenth Army for the remainder of the war.

Even as the Allies fought their way into Buna, a second successful campaign against the Japanese was being brought to its conclusion. In June 1942 Captain Martin Clemens, a coastwatcher stationed on Guadalcanal, Solomons Islands, reported that the Japanese were constructing a major airfield near Lunga Point. The implications were extremely serious, since bombers stationed there could interdict the supply route between the United States and Australia as well as provide cover for strikes against American bases in the New Hebrides to the south-east. The American Joint Chiefs of Staff were already planning to take the offensive in this area and it was decided that the airfield and the nearby island of Tulagi would be captured by Major-General Alexander A. Vandergrift's reinforced 1st Marine Division.

The Marines landed on 7 August, driving the enemy's construction troops off the airfield, which was named Henderson Field in honour of Major Lofton Henderson, a Marine officer who had been killed during the Battle of Midway. The airfield was completed and on 20 July the first aircraft of what became known as the Cactus Air Force began touching down. To ease congestion a fighter airstrip, designated Fighter One, was constructed 2,000 yards east of the main airfield and became operational on 9 September.

The Japanese reaction was immediate and violent. Using fast destroyer transports, they landed troops at Taivu Point, east of the airfield, and at Kokumbona, to the west, continuing to build up their strength until, by mid-October, they had over 20,000 men ashore, commanded by Major-General Haruyoshi Hyakutake and designated the Seventeenth Army. Simultaneously, they launched naval and air offensives against the American beach-head. These resulted in no less than seven major surface engagements in which the Japanese showed a marked superiority in night fighting techniques which, coupled with their formidable Long Lance torpedoes, enabled them to emerge the victors on most occasions, although they were unable to retain control of the sea. The Allied navies lost 24 major warships and the Imperial Navy 18, but for the Japanese these losses were irreplaceable, while those of the Americans could be absorbed by their huge ship-building programme. Air activity was almost continuous, the fighters from Henderson Field taking a heavy toll of the enemy's bombers. The bomber crews, flying from distant bases, must have wondered why their opponents were always waiting up-sun, ready to pounce; the answer was that coastwatchers on islands further along the Solomon chain regularly reported their passage in plenty of time for the fighters to be scrambled. In addition to being subjected to air attack, Henderson Field was often shelled from the sea. During the night of 13/14 October the battleships *Kongo* and *Haruna* landed 918 14-inch shells, plus thousands of

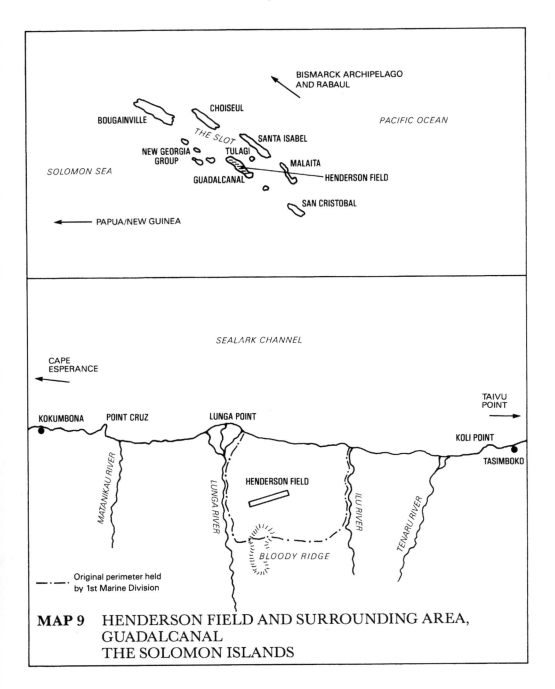

MAP 9   HENDERSON FIELD AND SURROUNDING AREA, GUADALCANAL THE SOLOMON ISLANDS

Above *Henderson Field, Guadalcanal, from the air. For both sides, the entire campaign revolved round possession of the base. Although the photograph was taken after the fighting had ended, the whole area is pockmarked with bomb and shell craters.* (US Marine Corps)

Below *Anti-aircraft gun crew on Rendova, New Georgia Group, Solomon Islands. The battery to which the gun belongs set something of a record by destroying 12 out of 16 enemy bombers for the expenditure of only 88 rounds — its 'kills' being painted on the barrel.* (USMC)

lesser calibre projectiles, on or around the airstrips. Despite this, there were few occasions when the field was not operational. Throughout the campaign, both sides continued to reinforce and resupply their troops, the Americans by day under the cover provided by the Cactus Air Force, the Japanese by night using the so-called Tokyo Express, which consisted of fast destroyers and transports.

Critical to the outcome of the campaign as the naval and air operations were, we are concerned here with the battles fought by the ground troops, battles fought in a climate and in jungles which were, if possible, every bit as bad as those in New Guinea. For both commanders the primary objectives were very simple. Vandergrift had to hold Henderson Field and prevent the Japanese interfering with its operation; for his part, Hyakutake intended recapturing it.

On 19 August marked maps captured from an ambushed Japanese patrol east of the perimeter indicated that the force landed at Taivu, part of the 28th Regiment under the command of Colonel Kiyanao Ichiki, intended attacking the airfield. Ichiki had a poor opinion of his opponents and, rather than await the arrival of the rest of his regiment, decided to advance with the 900 men already ashore. The loss of his patrol does not seem to have worried him unduly, despite the fact that he knew very little about the strength of the American defences. At 02:40 on 21 August he led 500 of his soldiers in a wild charge across the sandbank at the mouth of the Ilu River, sometimes referred to incorrectly as the Tenaru, which lay three miles to the east. Vandergrift had reinforced this sector and, although a few men reached the American wire, the remainder were cut down by concentrated fire. At 05:00 Ichiki tried again, this time along the shore, with similar results. The Japanese retired to a

A photograph which illustrates the difficulty of moving through coastal mangrove swamp. (USAMHI)

coconut plantation east of the river, where they were subjected to artillery fire and strafed from the air throughout the morning. Meanwhile, Vandergrift had ordered the I/1st Marines to cross the Ilu upstream and attack the enemy's flank and rear. At 15:00, while this attack was taking place, five Stuart tanks of Company A, 1st Marine Tank Battalion, crossed the body-strewn sandbank to deliver the *coup de grâce*; afterwards, Vandergrift was to comment that the tank tracks resembled 'meat grinders'. Only Ichiki, one other officer and a handful of survivors returned to Taivu; there, the regimental colours were burned and Ichiki took his own life. The Battle of the Ilu cost the Marines 35 dead and some 50 wounded.

Next it was the turn of Major-General Kiyotake Kawaguchi, who had at his disposal the second echelon of Ichiki's regiment and most of the 124th Regiment. Kawaguchi's intention was to assault Vander-grift's southern perimeter under cover of feint attacks along the Ilu, and to this end he set off from Taivu on 6 September, his men slopping through the rain-drenched jungle in an untidy column three miles long. Unfortunately for him, on 8 September an American amphibious raid on the village of Tasimboko, executed by Colonel Merritt Edson's 1st Raider Battalion, the Marine equivalent of the British Commandos, secured documents containing details of his plans. Once again, Vandergrift was able to take precautionary measures. The combined 1st Raider and 1st Parachute Battalions, under Edson's command, plus additional artillery, were moved to the threatened sector, which subsequently became known as Bloody Ridge.

Kawaguchi's troops began reaching the area during the evening of 12 September, but instead of waiting for them all to arrive and allowing them a short rest period, he

A Stuart gives covering fire to infantry pinned down on a jungle track. Bougainville, Solomon Islands, 1944. (USAMHI)

launched a series of piecemeal attacks which were easily defeated. The following night the assault was renewed in strength, covered by a naval bombardment. Despite the fact that the combined fire of the American infantry and artillery was blowing great holes in the packed Japanese ranks, the defenders were forced to yield some ground and at one stage Vandergrift's escort and clerks killed three of the enemy who penetrated as far as his command post. By dawn, however, the Japanese had melted back into the jungle, leaving behind them over 700 dead. Kawaguchi embarked on a withdrawal to the west, abandoning his heavy weapons as he went, the eight-day retreat being rendered the more difficult by the need to carry over 500 wounded. American casualties at Bloody Ridge amounted to 59 killed and 204 wounded.

During the second half of September and early October Vandergrift mounted a series of operations designed to extend his western perimeter beyond the Matanikau River. These depended too heavily on separate units being able to co-ordinate their movements, which was hardly possible given the nature of the terrain, and resulted in serious casualties. The Japanese, who had themselves been concentrating for an attack across the Matanikau, also suffered severely; on 9 October their 4th Regiment sustained over 600 killed when it attacked a ridge along which the I/7th Marines were withdrawing to the perimeter.

Meanwhile, the Japanese build-up continued and Hyakutake planned a major counter-offensive. Using a trail cut by engineers from Kokumbona to a point south of the American perimeter, Lieutenant-General Masao Maruyama's 2nd Division would split into two wings, the right commanded by Kawaguchi and the left by Major-General Yumio Nasu. These would then overrun the Marines' defences and penetrate as far as the sea, capturing Henderson Field in the process. Simultaneously, a 3,000-strong force under Major-General Tadashi Sumiyoshi, reinforced with a heavy artillery group and a tank company, would make

diversionary attacks across the Matanikau.

Maruyama began his march along the newly-cut trail from Kokumbona on 16 October. Progress was painfully slow, the best that the troops could manage being six miles a day, and that at the cost of abandoning the heavier artillery weapons. The assault date was deferred from 18 to 22 October, then to 23 October, and finally, because Kawaguchi was still cutting his way through the jungle, to 24 October. Furious, Maruyama dismissed the unfortunate general and replaced him with Colonel Toshinaro Shoji.

Incredibly, Sumiyoshi was not informed of this final postponement and, as planned, he began his diversionary attack across the mouth of the Matanikau at sunset on 23 October. The Americans, already alerted by a probing attack and increased artillery activity, were waiting for him. His tanks were smashed to scrap-iron by concentrated artillery and anti-tank gun fire as they attempted to cross the sandbank, and 650 of his men were killed trying to advance without their support.

By now, Maruyama's columns had also been spotted. When they attacked during the night of 24/25 October the battle followed a familiar pattern. Heedless of their casualties, the Japanese advanced with suicidal bravery, the weight of their assault falling on Lieutenant-Colonel Lewis Puller's I/7th Marines. Amid ferocious hand-to-hand fighting the Marines slowly gave ground but at 02:00 the III/164th Infantry Regiment, an Army unit, was committed to the fight and the line held. At first light the Japanese retired into the jungle, leaving bodies piled on the slopes fronting the American positions; among them was that of General Nasu. The next night a further assault was mounted on the upper reaches of the Matanikau but this was thrown back by a scratch force of signallers, clerks, cooks and drivers.

While the offensive cost the Americans some 200 killed and a similar number of wounded, Maruyama's losses amounted to over 2,000 in dead alone. He despatched

Left *Combat photography in the jungle was no picnic. The cameraman on the right was wounded by shrapnel only minutes after taking this shot of the US 25th Division in action at Balete Pass, Luzon, on 24 March 1945.* (USAMHI)

Right *Machine gun post of the US 37th Infantry Division, Laruma River, Bougainville, March 1944. Vision seems restricted, but the vegetation will be quickly cut away when the weapon opens up.* (USAMHI)

Left *4.2-inch mortar in action against Japanese positions on Arundel Island, New Georgia Group, September 1943.* (USAMHI)

Right *Japanese bunkers were able to withstand heavy shellfire but at close quarters could be silenced by manpack flame-throwers. New Georgia, August 1943.* (USAMHI)

some of the survivors under Shoji to march east to Koli Point and withdrew with the rest to Kokumbona. Vandergrift immediately began expanding his perimeter with the object of forcing the enemy to withdraw their heavy guns, known to the Americans as Pistol Petes, beyond range of Henderson Field. A bridgehead was secured west of the Matanikau but on 1 November operations were suspended when it was learned reinforcements for Shoji were being landed at Koli Point. This development, it was thought, heralded the start of a fresh Japanese offensive, although Hyakutake was simply building up Shoji's strength to the point at which he would be able to march round the American perimeter and rejoin the main body of the Seventeenth Army west of the Matanikau. On 3 November the II/7th Marines, operating east of the perimeter, were first subjected to frontal attacks at the Metapona River and then isolated when the Japanese moved round their inland flank. Vandergrift immediately despatched the I/7th Marines, III/164th Infantry and sup-

porting artillery to the assistance of the trapped battalion, under the overall command of Brigadier-General E. B. Sebree. By 10 November Sebree was in position to attack and the Japanese were in turn trapped between his troops and the II/7th Marines, holding their ground at Gavaga Creek. Some 350 Japanese were killed but, led by Shoji, a larger party broke through the 164th Infantry and escaped inland, having abandoned their artillery and most of their rations.

In pursuit Vandergrift sent Lieutenant-Colonel Evans Carlson's 2nd Raider Battalion, accompanied by native guides and porters. Carlson's tactics were simple. His main body moved parallel to Shoji's, detaching snipers to pick off the enemy's officers and senior NCOs; simultaneously, a fighting patrol would harry the Japanese rearguard until reinforcements were sent back to help; these would then be ambushed and destroyed by the rest of the battalion. In this way the Raiders savaged the remnants of Shoji's command on no less than twelve

separate occasions, killing over 500 Japanese for the loss of only 17 of their own men.

For Hyakutake the position soon became desperate. He had plenty of troops ashore, but many were wounded and more were riddled with disease. He was critically short of ammunition, food and supplies of every kind and, as a result of American naval and air activity, very little was being landed. By way of contrast, the well-supplied Americans had renewed their outward pressure and in early December Vandergrift and his tired 1st Marine Division were relieved by Major-General Alexander Patch and the leading elements of US XIV Corps, which included the Americal, 25th and 2nd Marine Divisions.

It was now apparent, even in Tokyo, that the cost of holding Guadalcanal was prohibitive. General Hideki Tojo, the Prime Minister, had accorded Guadalcanal a higher priority than New Guinea and stubbornly refused to consider withdrawal. Since November, however, Yamamoto had insisted that the Imperial Navy could not sustain a continued rate of attrition which, in addition to the warships sunk or damaged, had cost Japan over 600 experienced naval pilots and 300,000 tons of merchant shipping sent to the bottom. His view prevailed and on 31 December the Emperor consented to the evacuation of the Japanese garrison. For the sake of security the order was delivered to Hyakutake by the Tokyo Express during the night of 14/15 January 1942.

Patch, of course, knew nothing of this and had gone over to the offensive on 10 January. His troops made slow progress against well-entrenched, determined opposition but after the evacuation order was received the Japanese fell back doggedly to Cape Esperance, from which the last 11,000 men of the Seventeenth Army were taken off by the Tokyo Express between 1 and 7 February. The Japanese deception plan succeeded brilliantly, so that when Patch's troops launched their final assaults during 8 and 9 February, they found only abandoned positions and wrecked equipment. It was galling for the Americans to be balked of their prey,

but as Admiral Chester Nimitz was to comment: 'Until almost the last moment it appeared that the Japanese were attempting a major reinforcement effort. Only skill in keeping their plans disguised and bold celerity in carrying them out enabled the Japanese to withdraw the remnants of the garrison. Not until after all organized forces had been evacuated did we realise the purpose of their air and naval dispositions; otherwise, with the strong forces available to us on Guadalcanal and our powerful fleet in the South Pacific, we might have converted the withdrawal into a disastrous rout.'

The Allies were to win further victories as they extended their advance along the Solomons chain, but none of these campaigns were as critical to the fortunes of either side, nor were they fought with the same bitter intensity, as had been that on Guadalcanal.

For the British, real success against the Japanese was longer in coming. In December 1942 it was decided to mount a strictly limited offensive which would restore morale. The area selected was the Arakan

coast of Burma. This consists of mangrove swamps and twisting tidal chaungs, while inland the steep-sided ridges which separate the river valleys are clothed in almost impenetrable jungle. The Arakan is exposed to the violent electrical storms generated by the Bay of Bengal and between May and September is subject to a rainfall of 200 inches, which can wash out the primitive tracks in a day, although for the rest of the year it is dry and dusty. For the Japanese, however, the Arakan was of considerable strategic importance, since it provided a route into the vulnerable regions of Central Burma, and for this reason it was held in strength.

The British offensive was to be carried out by Major-General Lloyd's 14th Indian Division and its object was to secure the Mayu Peninsula. For a while all went well, but in January 1943 the advance was halted by an extremely strong Japanese position at Donbaik, ten miles short of Foul Point, where the defences were constructed in depth. After several costly attacks failed to break through, eight Valentine infantry tanks belonging to 146 Regiment RAC were

Above left *A Sherman grinds forward at first light to assist infantry whose positions have been infiltrated by the Japanese during the night. Bougainville, March 1944.* (USAMHI)

Above *A Marine Corps 155mm gun is emplaced on Rendova Island, July 1943.* (USAMHI)

Right *Once in position, the gun was used to bombard Munda airstrip on New Georgia Island, 11 miles across the sea.* (USAMHI)

Above *A Matilda of the Australian 2/4th Armoured Regiment crosses the Puriata River, Bougainville, March 1945. The wire grille over the engine deck is a defence against hand-placed explosive charges and was also used by British tanks in Burma.* (Australian War Memorial)

Left *General Douglas MacArthur with Australian and American officers during a tour of inspection at Balikpapan, Borneo, July 1945.* (AWM)

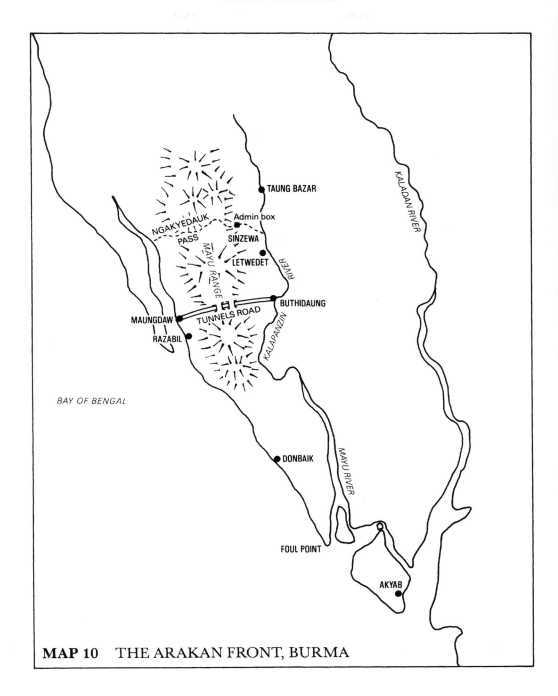

MAP 10 THE ARAKAN FRONT, BURMA

During the Guadalcanal campaign Amtracs, otherwise known as LVTs (Landing Vehicles, Tracked), were used for ship-to-shore logistic support. (USAMHI)

brought forward and committed in support of a Dogra battalion on 31 January. The Valentine was well armoured and extremely reliable, but the commander had little vision when closed down and since the driver relied upon him for direction across difficult going, it was only too easy to drive into unseen obstacles. The half-squadron's two troops reached the enemy bunker complex but, shortly after, three tanks became bogged in a ditch. The remainder managed to return to the start line, although one was penetrated by an anti-tank gun, killing the gunner, and another had to be towed in under fire after it had broken down. The stalemate at Donbaik continued.

Meanwhile, the Japanese 55th Division, commanded by Lieutenant-General Takishi Koga, was preparing a counterstroke. In March it advanced from the Kaladan valley and crossed the Mayu Range, hoping to sever 14th Indian Division's communications and trap the considerable British force at Donbaik. Lloyd was compelled to execute a difficult withdrawal, assisted by the RAF and two reinforcement brigades which Slim pushed forward, but by 12 May the British were back where they had started in December. Far from restoring morale, the offensive actually lowered it, since it seemed to indicate that the Japanese were still the masters of jungle warfare.

Elsewhere, however, the Japanese in Burma had received a most unpleasant surprise. In 1941 Brigadier Orde Wingate had commanded a group known as Gideon Force which had operated successfully behind Italian lines during the campaign in East Africa, tying down no less than seven brigades that were badly needed elsewhere. Wingate believed that a similar group, supplied by air, was capable of operating for protracted periods behind the Japanese lines, where it could cut communications,

74

In subsequent amphibious operations against Japanese-held islands Amtracs were employed in the assault role. Here, Marines train in their use on Guadalcanal. (USMC)

destroy supplies and generally inflict damage out of proportion to the number of troops employed. Wavell approved the idea and Wingate was given permission to raise an experimental force. Officially, this was designated 77th Indian Brigade, but became better known as the Chindits from its distinctive arm badge, a *Chinthe* or stone lion which guarded the entrance to Burmese pagodas.

Using animal transport to carry heavy weapons, ammunition, supplies and radio sets, the Chindits crossed the Chindwin on 18 February 1943, organised into seven columns each 400 men and 100 mules strong. The Mandalay-Myitkyina railway was destroyed in several places and a number of successful actions were fought against the enemy. In March Wingate led his men across the Irrawaddy but by now he had stirred up a hornets' nest, for the Japanese hated having their communications inter-

fered with as much as they loved interfering with other people's. The Chindits were forced to embark on a nightmare retreat, suffering from disease, exhaustion, lack of water and food. By the time they reached India in April they had marched an average of 1,500 miles through territory that was nominally controlled by the enemy, a fact which did much to restore British confidence and offset the disappointing results of the Arakan offensive. Much, too, had been learned in the fields of close fighter support, air supply and casualty evacuation, but the price had been a heavy one. Of the 2,182 survivors, only 600 were considered fit for further active service, although many of these formed the nucleus of the greatly enlarged Chindit force which would take the field the following year.

In the meantime, the reasons for the failure in the Arakan were being analysed at General Headquarters in Delhi. Astonishing as it may seem, there were still senior staff officers who, despite the object lessons taught by the Japanese in Malaya and 7th Armoured Brigade in Burma, clung to the view that tanks had little place in jungle warfare and pointed to the débâcle at Donbaik to reinforce their views. Some even went so far as to tour British and Indian armoured units, informing their officers that if they wished to see anything of the war they had better transfer to the infantry, as tanks would never be used in Burma. Fortunately, the hour produced the man in the person of Brigadier Reginald Scoones, the commander of 254th Indian Tank Brigade, who had just returned from Imphal, where his brother, Lieutenant-General G. A. P. Scoones, commanded IV Corps. The brigadier had surveyed the corps' two principal axes, the Tamu and Tiddim roads, and was convinced that tanks could operate along these, albeit on a limited frontage. Now, with the disbandment of his brigade a very real possibility, Scoones felt that he had nothing to lose and flew to Delhi, where he expressed himself forcefully. He was asked by one senior officer how long he had been in India. 'I told him six months'. He replied, 'We have been here many years and we know India', to which I replied, 'Well at least I have walked the course on my own feet, which is more than anyone in this Headquarters has done!' Scoones persisted and eventually reached General Sir Claude Auchinleck, the Commander-in-Chief India since Wavell's appointment as Viceroy. He pointed out as strongly as he could that tanks *could* be used in Burma, would eventually *have* to be used, and that if they were not used the army would be seriously caught short. He won his point and 254th Tank Brigade was sent forward to Imphal.

A new offensive was planned in the Arakan, where the Japanese had halted their advance along the line Maungdaw-Buthidaung, two towns on either side of the Mayu Range, connected by the only metalled all-weather road in the area. This ran along the old track-bed of a disused narrow-gauge railway which, at its summit, ran through two tunnels. The capture of the road, together with Maungdaw and Buthidaung, was the

Left *Fire and movement. British infantry fight their way methodically through a banana plantation in the Arakan. Note the muzzle of the Bren light machine gun giving covering fire as the troops on the right move forward.* (IWM)

Right *Mahratta infantry prepare to attack through a bamboo thicket. Arakan, 1944.* (IWM)

object of the offensive, for without it troops east of the Mayu Range could only be supplied along a winding fair-weather track running through the Ngakyedauk Pass some miles to the north. Once the Tunnels Road and its termini had been secured, further operations could be planned.

The offensive would be carried out by Lieutenant-General A. F. P. Christison's XV Corps, consisting of the veteran 5th Indian Division under Major-General H. R. Briggs, with a long list of hard-fought actions in East Africa and the Western Desert to its credit; the 7th Indian Division, commanded by Major-General F. W. Messervy, who had also served in the Western Desert; two brigades of Major-General C. G. Woolner's 81st West African Division, containing many men whose home was the jungle and whose ancestors had fought against Wolseley; and, in reserve, 26th Indian Division (Major-General C. E. N. Lomax), which had held the line following the withdrawal from the Arakan earlier in the year.

Also at Christison's disposal was an armoured regiment, the 25th Dragoons

under Lieutenant-Colonel H. C. R. Frink, equipped with M3 Lee medium tanks. Although the Lee was regarded as obsolete in other theatres of war, its configuration was ideal for jungle warfare since on narrow paths its sponson-mounted 75 mm gun was able to engage targets ahead while the 37 mm in the turret could fire to either flank, often using deadly man-killing canister ammunition. During the First Burma Campaign the Japanese infantry had shown little fear of tanks, which they attacked with petrol bombs, pole charges and other explosive devices, and for this reason the Dragoons— and indeed every British and Indian armoured regiment which served in Burma—were accompanied by a close escort of Bombay Grenadiers, whose battalions had been trained for this dangerous and exacting work, in which they excelled.

Christison's plan involved 5th Indian Division advancing down the coast to Maungdaw while 7th Indian Division moved on a parallel axis on the other side of the Mayu Range and the West Africans guarded the left flank. Maungdaw was captured on 9

January 1944 but 5th Division then found itself confronted by an extremely strong position at Razabil, a little distance beyond, and the 25th Dragoons were called forward. A bunker-busting technique was quickly evolved in which the tanks fired high explosive shells ahead of the infantry with gradual lifts until the scrub had been blown off and the target bunker hit. At this point the tanks would switch to armour-piercing shot, which smashed up the bunker timbers, and the infantry's mortars would open a heavy fire on the crest and reverse slopes of the feature, pinning the enemy down while the riflemen launched their assault with bayonet and grenade. The technique worked, although progress was slow and in due course the tanks' movements were inhibited by an area of soft ground and chaungs running between banks 20 feet in height. By the end of January a stalemate had been reached.

By this time air reconnaissance and reports from V Force, a small group of extremely brave officers and men who undertook special operations along the Arakan coast, indicated that substantial Japanese reinforcements were arriving through Akyab and that these would probably be used in a counter-offensive east of the Mayu Range. It was, therefore, vital that 25th Dragoons should cross the range and reach the Buthidaung plain in time to defeat this, and Christison ordered his engineers to turn the Ngakyedauk trail into a track capable of supporting tanks and lorries. On 4 February the Dragoons began moving through the pass, leaving their reserve tanks and crews with 5th Indian Division.

It was not a moment too soon. Lieutenant-General Hanaya, the new commander of the 55th Division, had planned a scaled-up version of the counter-offensive which had defeated the British incursion into the Arakan the previous year. The main

A 3.7-inch mountain howitzer battery of 7th Indian Division in action against Japanese positions in the Ngakyedauk Pass. (IWM)

body of the division, under Major-General Sakurai, was to infiltrate the gap between the 7th Indian and 81st West African Divisions, capturing Taung Bazar on the former's lines of communication. Simultaneously, a force under Colonel Kubo was to cross the Mayu Range and cut off supplies to the 5th Indian Division, while a third column under Colonel Tanahashi swung south and took the Ngakyedauk Pass. Hanaya anticipated that both Indian divisions, finding themselves isolated, would attempt to fight their way out and that the West Africans would conform to their withdrawal. In the longer term, he intended threatening Chittagong and was confident that to retrieve the situation Slim would have to strip the Imphal front at the very time the Japanese planned to launch a fresh offensive there. Unfortunately, in planning his own offensive, which he designated *Ha-Go*, Hanaya failed to allow for two very important factors. The first was that since the arrival of Spitfire squadrons in December, it was the RAF which ruled the Arakan skies, and this in turn meant that isolated troops could be supplied by air-drop. The second was the presence of Frink's tanks.

Even as the Dragoons emerged from the Ngakyedauk Pass it was clear that large-scale infiltration was taking place. On 6 February Messervy's headquarters was overrun, although the general and most of his staff managed to reach the defensive box held by 7th Division's administrative troops at Sinzewa. Two squadrons of Lees patrolled the road between the Admin Box and the former headquarters, covering the withdrawal of troops into the box, recovering the 5.5-inch howitzers of 6th Medium Regiment from the mud, escorting a mule train and recovering a number of vehicles which had been abandoned.

The Admin Box itself measured 1,500 yards from east to west and half that from north to south, being almost cut in two by a feature known as Artillery Hill. Into this space were packed tanks, guns, transport, tactical and administrative headquarters, supply dumps and a field hospital, so that every shell fired by the Japanese was almost certain to find a target, especially as the position was overlooked from every direction by hills varying in height from 100 to 200 feet. The perimeter was manned by artillerymen fighting as infantry and a scratch force of service troops, with 25th Dragoons, their Bombay Grenadier escort and two companies of the 2nd West Yorkshire Regiment acting as mobile reserve. While Messervy continued to run his division from an improvised command post, the defence of the box was the responsibility of Brigadier G. C. Evans, a no-nonsense soldier who bluntly informed his mixed bag of a garrison that they had two choices—they could fight like hell and keep the Japanese out, or captivity with the likely prospect of being butchered. He further instituted a daily competition to establish on which sector the highest total *per capita* of fresh Japanese bodies might be found.

The siege of the Admin Box lasted until 25 February. Tanahashi's artillery shelled the interior continuously and his infantry made frenzied attempts to break through the defences, being thrown back by the defenders and the point-blank fire of the Lees. At night, when the tanks were less effective, the Japanese were able to approach much closer before launching their assaults, and innumerable vicious little personal battles were fought. During the night of 7th February a mixed party of Japanese and Jifs (Indian prisoners, mostly captured in Malaya, who had thrown in their lot with the enemy) broke into the main dressing station, slaughtering doctors, orderlies and patients without mercy. They then dug in amid the shambles until ejected by a West Yorkshire company and troop of Lees two days later. Fifty Japanese were killed, including an

The interior of the Admin Box, showing several of 25th Dragoons' Lees and other vehicles. (IWM)

officer on whose body was found a complete set of plans for the *Ha-Go* offensive. The principal effect of the massacre, however, was to generate revulsion and anger. Hitherto, British and Indian troops had regarded the Japanese with a respect amounting almost to awe; now, at last, they understood the enemy's concept of extermination and were prepared to respond in kind.

Despite this incident, on 9 February Evans felt sufficiently secure to despatch the Dragoons' A Squadron to reinforce 33rd Indian Brigade's box some miles distant, the move being completed safely. On 11 February, supplies were dropped by air for the first time, and continued to be dropped regularly throughout the siege. For the Japanese, who had consumed what rations they had and been unable to capture those upon which they had relied, the sight of the parachutes drifting down was galling in the extreme.

Hanaya was becoming seriously concerned by his lack of progress. Instead of retreating, the British were stubbornly holding their ground and inflicting severe casualties on 55th Division. Furthermore, Christison had committed 26th Indian Division, which was advancing south and threatening to trap Kubo and Tanahashi between the hammer and the anvil. Simultaneously, the 123rd Brigade from 5th Indian Division, spearheaded by a composite Lee squadron formed from the Dragoons' spare tanks and crews, was pushing through the Ngakyedauk Pass towards the Admin Box. On 13 February the weary garrison was informed that relief was on the way and two days later the leading battalion of 26th Division made brief contact with the defenders before enemy pressure forced it to withdraw. In addition, several infantry companies from Messervy's own division managed to work their way into the box, taking some of the burden off the exhausted West Yorkshires.

On reaching the summit of the Ngakyedauk Pass 123rd Brigade found itself confronted by a large bunker complex. The bunkers were so well constructed that not even the fire of the Lees' 75 mm guns could suppress them and an unusual solution was adopted. A 5.5-inch howitzer was brought up and two tanks manoeuvred into position to provide protection for the gun and its crew. Suddenly aware of their danger, the Japanese fired everything they had at the group, but it was too late. Methodically, the howitzer slammed its heavy shells into one fire slit after another, and after 20 rounds had been fired all that remained of the complex was smoking craters. The advance continued and on 22 February contact was made with the garrison, and although the Japanese remained in the area for a further three days, the siege was as good as over. The interior of the Admin Box reeked of death and cordite, was covered with the skeletons of burned-out vehicles and smashed equipment, and the new dressing station contained 500 wounded men. Both the garrison and the relief force, however, sensed that they had won a critical victory.

Hanaya formally abandoned *Ha-Go* on 24 February. The offensive had cost 55th Division 5,600 killed and its survivors, ragged and starving, were forced to withdraw. Christison's corps sustained 3,500 casualties but was able to resume its advance.

Since the Second Battle of Alamein the British Army had become adept at controlling large numbers of guns, switching their fire around the battlefield by means of an efficient radio net. Thus, when 7th Indian Division attacked a heavily fortified area at Letwedet on 6 March, it was supported by the combined firepower of 75 25-pounders, 16 3.7-inch howitzers and 32 5.5-inch howitzers, which pulverised the defences and demoralised the Japanese. On 11 March Buthidaung itself was taken with tank support after a similar concentration landed 7,000 shells in the target area within the space of 15 minutes. Across the Mayu Range the defences of Razabil were stormed from the rear after 5th Division's 161 Brigade had made a wide flank march through the foothills. The assault went in shortly after first light on 12 March, supported by the fire of no less than 200 guns and dive-bombers which turned the formidable bunker complex into a tangle of splintered timber and torn sandbags. XV Corps had now been joined by Major-General Francis Festing's 36th Division, supported by the Shermans of C Squadron 149 Regiment RAC, and this completed the capture of those sections of the Tunnels Road still in enemy hands. In his book *Defeat into Victory* Slim recalls that on 27 March 'a Welsh battalion, supported by tanks, assaulted the defences of the Western Tunnel. In the mêlée a tank fired a shell directly into the tunnel mouth. Ammunition stored there blew up in a series of stunning explosions, and in the confusion the Welshmen rushed the enemy and the tunnel was ours.'

Christison had secured all his objectives, although reinforcements from the Japanese 54th Division were reaching the front and these mounted several determined counter-attacks until the onset of the monsoon in May put an end to the fighting. At last a tangible victory had been won and, far from stripping the Central Front to defend the Arakan, Slim was able to commit the 5th

81

The Western tunnel on the Maungdaw–Buthidaung road, severely damaged when ammunition stored inside was exploded by gunfire from Shermans of 149 Regiment RAC. Engineers have effected temporary repairs to reopen the road. (IWM)

and 7th Indian Divisions to the battles raging at Imphal and Kohima. For the British, learning the business of jungle warfare had been a painful process, but they were about to inflict a shattering defeat from which the Japanese Burma Area Army would never recover.

CHAPTER 4

'HONDA SPEAKS OF NOTHING ELSE!'

By the beginning of 1943 the Allies were evaluating the lessons learned during their first victories over the Japanese, and the myth of the latter's invincibility had begun to wear thin. The following, which first appeared in a US Marine Corps pamphlet, was reproduced by GHQ Delhi in a training memorandum of June 1943 and distributed to British and Indian troops destined for the Burma front:

It is true that the Japanese is a good fighting man, that he will fight to the death rather than surrender and that his offensive tactics are carried through with dash and spirit and frequently with fanatical courage, but it is also true that these very characteristics play into our hands and aid in destroying him. That we are fully capable of out-thinking, outshooting and outfighting the Japanese has been amply demonstrated by the Marine Corps and Army units who have fought and beaten the best jungle troops of Japan in the Guadalcanal, Tulagi and New Guinea areas.

Much has been said of the ability of the Japanese to slip silently and invisibly through the jungle, particularly at night. Experience has proven that the Japanese is anything but silent and anything but an outstanding bushman. He has frequently been caught by our own patrols in the densest jungle country as a result of the disclosure of his own whereabouts by constant chattering and conversations and the complete absence of the most elementary forms of local security. Here our own ability to advance stealthily and noiselessly to within a few yards of an unsuspecting enemy, to fire fast and accurately, has paid

large dividends at minimum cost. Avenues of approach into our positions are frequently prepared in advance, foxholes dug, and paths lined with vines to facilitate night movement. Constant patrolling to front and flanks quickly discloses these activities, and the judicious use of ambushes nullifies them.

The Japanese preference for night attacks is disturbing only if methods for combating them are not fully understood. Infiltration tactics through ravines are easily detected by alert cossack (i.e. listening) posts, noise devices and booby traps, and easily combated by proper placing of tactical wire, obstacles and bands of machine gun and mortar fire. Avenues of approach are easily covered by artillery fire.

The attack is presaged by much small arms and mortar fire and disclosed by a mass rush accompanied by loud *banzais* and frequently, taunts and misleading commands in English. Tactical wire properly placed and final protective lines of automatic fire, plus a determination to hold one's ground, reduces the attack to a shambles and results in a completely destroyed enemy.

The frequent stress laid on the Japanese ability to operate in the jungle for long periods of time on little food and his ability to endure physical hardship will bear closer examination. Our troops, with their minds filled with such stories, should understand that the Japanese is no better able to go without food than they are, that their stamina is no better than our own, provided the necessary steps have been taken to insure top physical condition. The Japanese gets just as wet when it rains and suffers just as much, if not more, from malaria, dysentery, dengue, ringworm and other forms of tropical ills. This has been amply born out by the condition of prisoners captured and by finding dead who literally starved to death.

The British, too, were studying the nature of jungle warfare and established a Jungle Warfare Training Centre at Gudalore in the Nilgari Hills, where battalions moving up to the front underwent a three-week course. Here, in addition to being taught such basic skills as bedding down, soldiers were instructed in elementary bushcraft techniques in a commonsense manner which would have been familiar to Robert Rogers and his Rangers. (See Chapter 1)

No lights of any kind will be used in forward positions during hours of darkness and there will be no smoking unless specifically authorised.

Silence is essential at the front and on patrol. If the enemy can't locate you, every advantage is with you. He will try by every means to get you to disclose your position so that he can gain the advantage over you.

Don't shoot unless you have something worthwhile to shoot at. Blind shooting simply gives away your location and may kill your own comrades.

Be on guard for all types of booby traps and other enemy ruses.

Japanese light mortar fire is accurate; therefore bunching of personnel must be strictly avoided, and individuals will not allow themselves to be silhouetted on the skyline.

Dig in whenever halted and improve holes as time permits.

Talk only in as low a tone as possible. Practise whispering, especially in telephone conversations. Use signals, such as hand or arm signals, tapping on the rifle, bird calls, etc, as much as possible. Do not expose yourself any more than necessary when using hand or arm signals. Remember that the careless soldier, who unnecessarily exposes himself, jeopardizes the security and lives of his team mates.

Be on the lookout for false surrender; any offer of surrender must be suspected as an enemy ruse.

Do not forget that many Japanese speak English. At night a favourite trick is for them to infiltrate and yell false orders such as 'Withdraw', or inquiries such as 'Sergeant, where are you?' Learn to know the voice of your leader. Japanese have difficulty in pronouncing the letter L.

Do not attempt to retrieve enemy wounded—there may be a trick.

Do not forget to look up before you move. Get in the habit of watching tops of trees as well as their roots.

When fired on by snipers, move at top speed to the nearest cover or concealment. As soon as possible, quietly change your position, then locate and destroy the sniper. Do not, in any circumstances, stand still in the open.

When challenging at night the sentry should remain concealed. Never permit an unknown person to come within knife range.

As the men's individual training progressed, they were familiarised with the jungle itself, taught how to cook their high-protein K-rations on small spirit stoves, follow game trails, preserve their sense of direction and interpret the distinctive tracks left by the enemy. They were also taught never to draw water downstream from a native village, for obvious health reasons, and to enforce strict water discipline when moving through dry areas, where thirst could induce unbearable torment; a former Chindit officer recalls that on one occasion he actually considered murdering his best friend for the contents of his water bottle, until a fresh source was discovered. Much careful thought had also been given to preventive medicine, so that in addition to receiving a number of inoculations the men were issued with anti-malaria mapacrine pills and salt tablets to ward off the worst effects of dehydration. It was recommended that as well as field dressings, soldiers should carry a small bottle of iodine to prevent cuts and insect bites turning septic, and a little permanganate of potash which was useful for sterilising water, disinfecting wounds and snake bite. The wearing of shorts was forbidden and a light mosquito net, dyed green, was considered essential when not moving or working.

After personal training had been completed, the men went on to learn the techniques of patrolling, ambush and counter-ambush, village clearing and working with other arms. As a result of accumulated experience, infantry battalions contained a higher proportion of automatic weapons—Brens, Stens and Thompson sub-machine guns—than in other theatres of war, and carried a larger number of grenades per man. After the Battle of the Admin Box, this training was also extended to administrative and line of communication troops.

Wingate, now a major-general, expanded his Chindit force steadily until it contained 77th Brigade, 111th Indian Infantry Brigade, a West African Brigade and three brigades formed from the 70th Division, recently arrived from the Middle East. The problem now arose of how to choose the best of these men for the gruelling task which lay ahead. Since stamina was an obvious requirement, this was solved by sending everyone on a 100-mile route march through the roughest jungle terrain possible and discarding those unable to keep up. Most Chindit recruits had served as conventional soldiers and found that their change of role to that of clandestine jungle raiders required some adjustment. They were, for example, taught to clear their camp sites so thoroughly before moving on that no trace of their presence remained for subsequent examination by Japanese patrols; this involved burying human and animal waste, ration packs and even cigarette ends, the in-fill itself being concealed with leaves and other forest debris.

On 14 October 1943 the British formations in Burma were designated the Fourteenth Army. Its commander, Lieutenant-General William Slim, had the gift of being able to establish a rapport with his troops, and his personal hold over them has been

A Chindit column crosses a dry river bed. Extra ammunition, supplies and radio sets were carried by mules. (IWM)

compared to that of Montgomery over the Eighth Army, although his quiet, gruff manner was very different. The victory in the Arakan had yet to come and he knew only too well that his first priority was to restore morale. This was accomplished partly by instituting the sound training measures described above, and partly by personal visits to units during which he listened patiently to what he was told. The principal complaint, among the British troops at least, was that, families apart, no one at home seemed remotely interested in what they were doing. The malaise was deep-seated and was not one to be cured by sympathetic noises. Shortly after, Vice Admiral Lord Louis Mountbatten, the Allied Supreme Commander in South-East Asia, also began visiting units and he did not mince his words: 'I hear you call this the Forgotten Front. I hear you call yourselves the Forgotten Army,' he would say. 'Well, let me tell you that this is not the Forgotten Front, and you are not the Forgotten Army. In fact nobody has even heard of you!'

A stunned silence was usually followed by angry muttering until Mountbatten continued: 'But they *will* hear of you very soon when you start beating the Japs—*and they will never forget you!*'

By the beginning of 1944 British and Indian troops understood Japanese methods, were familiar with the jungle and were fully trained. They were a very different proposition to the newly raised and inexperienced formations which had fought the battles of 1941 and 1942, and they believed that they could win.

The Japanese, however, remained firmly embedded in the attitudes and ideas which had produced their early victories. At the higher levels they resembled the Bourbons, who forgot nothing because they had learned nothing. The fact that their air and sea power was in decline and that they were now confronted by a fully mobilized enemy

with ample resources at his disposal does not seem to have influenced their belief that the old ways would suffice, nor to have caused them to review the less satisfactory aspects of those ways, notably their logistic infrastructure.

Nonetheless, the general situation in Burma, now held by an army group, the Burma Area Army, commanded by Lieutenant-General Shozo Kawabe, was one which gave grounds for concern. Apart from the fighting in the Arakan, Scoones' IV Corps at Imphal was growing stronger by the month and in the north Stilwell's Chinese army had to be contained. Resources were stretched to the limit and would clearly not suffice in the event of a major Allied offensive. This had been foreseen and in May 1942 it had been decided to build a railway from Bangkok in Thailand to Moulmein, where it would join the Burmese railway system, the object being to shorten lines of communication by eliminating the long sea passage to Rangoon around the Malayan peninsula. Using prisoners of war captured in Singapore, supplemented by an army of native labourers, the line was constructed from both ends with little more than hand tools and the occasional help of elephants. In addition to clearing the jungle along the route, it was necessary to excavate cuttings, build embankments and construct numerous bridges. The southern part of the route followed the Kwai valley but before this could be entered it was necessary to bridge the Mekhong River near Kanchanaburi. The first bridge, a timber structure, proved unequal to the task and was replaced by one of steel and concrete which was later bombed. As a result of the first Arakan campaign and the Chindit raids, construction was accelerated and the condition of the prisoners, already intolerable, degenerated even further. When the railway was officially opened in November 1943 it had cost the lives of 13,000 Allied prisoners of war and

90,000 native labourers, an average of 400 deaths for every one of its 250 miles, caused by disease, overwork, accidents and the brutality of the guards. Very few trains ran its full length, for Allied air attack rendered it virtually unworkable.

The first Chindit expedition emphasised the vulnerability of Japanese communications and when Stilwell opened a limited offensive in the Hukawng valley in October 1943, aimed in the long term at restoring the land link with China, Kawabe was further unsettled. He ordered Mutaguchi, now commanding the Fifteenth Army, to plan an offensive with the object of isolating and destroying Scoones' IV Corps at Imphal, following which the Japanese would establish an impregnable defence line along the crest of the Naga Hills, thereby thwarting any hopes the British might entertain of reconquering Burma from the north. Mutaguchi's plan, codenamed *U-Go*, involved the Fifteenth's Army's three divisions crossing the Chindwin on a broad front. Thereafter, the 31st Division, under Lieutenant-General Sato, was to sever Scoones' line of communication by cutting the road at Kohima; simultaneously, Lieutenanat-General Yamauchi's 15th Division was to attack the Imphal Plain from the north and east, pressing IV Corps back against Lieutenant-General Yanagida's 33rd Division, which would be closing in from the south and west. This all looked very well on the map, and might have worked during the early months of the war, but such was the difficult nature of the country and so great were the distances involved that cooperation between Mutaguchi's formations was impossible. Furthermore, so confident were the Japanese that once again they would be able to feed themselves from captured British stocks that the only additional provision made by the staff was for herds of cattle to be driven along behind the marching columns.

The situation was further complicated by the fact that Slim was himself preparing to take the offensive. Chiang Kai-shek had complained that if the Allies did not make greater efforts China would withdraw from the war, and since the effect of this would be to release 25 Japanese divisions for service elsewhere something had to be done. It was decided, therefore, that Stilwell would continue his advance down the Hukawng valley to effect the capture of Myitkyina and Mogaung, aided by Wingate's Chindits, who would be air-landed on top of the Japanese supply lines, while Scoones' IV Corps would advance to the Chindwin and effect a crossing. When Mutaguchi's intentions became clear Slim modified his plan. Stilwell's advance and the Chindit operation would proceed, but Scoones would concentrate his divisions on the Imphal Plain, where they could be supplied by air. Here the British armour could be used to best effect, and here the Fifteenth Army would be allowed to batter itself to pieces. When the Japanese had exhausted themselves, IV Corps would counter-attack.

On 5 March 1944 the Chindit airlift to north-central Burma began. Gliders were used to land the first wave in previously selected jungle clearings, which were rapidly fortified and transformed into airstrips capable of handling transport aircraft. The columns then fanned out to commence their work of ambush, destruction and harassment. One brigade for whom there was no air transport—Brigadier Bernard Ferguson's 16th—started early and marched every step of the way from Ledo in the Brahmaputra valley to its operational area near Indaw, a distance of 360 jungle miles, provoking the comment from one of its members: 'Ye gods, is all of Burma uphill?' Tragically, Wingate did not live to see his concept vindicated, for on 25 March his aircraft crashed into a mountain. In his place, Major-General Walter Lentaigne was appointed commander of the Chindit force.

Scoones' IV Corps consisted of Major-General Cowan's 17th Indian Division, operating in the Tiddim area; Major-General Gracey's 20th Indian Division in the Kabaw valley to the east; and Major-General Roberts' 23rd Indian Division in reserve at Imphal. In addition, two of 254th Indian Tank Brigade's regiments were available: the 3rd Carabiniers, supplemented by a squadron from 150 Regiment RAC, with Lees, and 7th Light Cavalry, the first Indian armoured regiment to take the field, with Stuarts. The corps' forward deployment stemmed from Slim's offensive intentions and the result was that when Mutaguchi's troops began crossing the Chindwin on 7 March clashes quickly ensued. One of the more spectacular of these took place near Tamu on 20 March when six Type 95 light tanks ambushed an equal number of Carabinier Lees from the flank on a jungle track. One Lee was set ablaze but the remaining five headed into a clearing, where they swung round and responded with their entire armament. After a brief duel the Japanese panicked and attempted to escape along the track. All six Type 95s were knocked out, one which did not burn being repaired and sent back to Imphal as a trophy.

Amid heavy fighting the 17th and 20th Indian Divisions completed their withdrawal to the Imphal Plain, inflicting serious casualties on the Japanese. The retreat of 17th Division was complicated by road-blocks, but Scoones despatched reinforcements, including A Squadron 7th Light Cavalry, and Cowan was able to fight his way out. Isobe Takuo, one of very few junior officers serving with 33rd Division to survive both the Imphal and Irrawaddy campaigns, recalls that in these preliminary encounters his 215th Infantry Regiment lost half its strength before the real battle had begun. Nevertheless, the pace of Mutaguchi's advance was alarming and between 19 and 29 March 5th Indian Division was flown into Imphal from the Arakan, less one brigade which was sent to Dimapur. On the 29th the road north from Imphal was cut by Yamauchi's men and a few days later Sato invested Kohima, leaving Scoones' corps completely isolated.

The Imphal Plain covers an area of some 700 square miles and lies 2,500 feet above sea level. It is entirely surrounded by

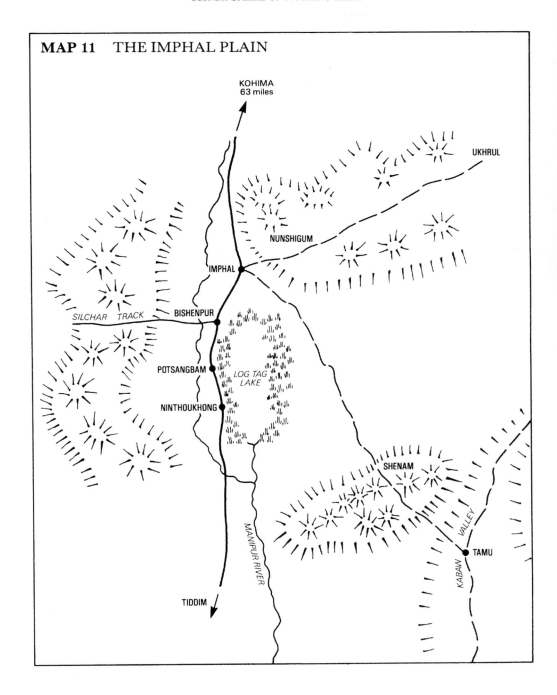

MAP 11 THE IMPHAL PLAIN

KOHIMA
63 miles

UKHRUL

NUNSHIGUM

IMPHAL

SILCHAR TRACK

BISHENPUR

LOG TAG LAKE

POTSANGBAM

NINTHOUKHONG

SHENAM

TAMU

KABAW VALLEY

MANIPUR RIVER

TIDDIM

scrub and jungle-covered mountains rising to a height of 8,000 feet. Through it, from north to south, runs the Manipur River, draining a marshy area known as Log Tag Lake at the southern end of the plain. The only reasonable road is that from the railhead of Dimapur, which enters from the north after passing through Kohima and makes its exit southwards through the mountains to Tiddim. Thus, if Imphal town lay at the centre of a clock, the Dimapur road would lie at twelve. At one is a massive detached feature known as Nunshigum. At two, some thirty miles distant, is the Ukhrul saddle, giving access by primitive jungle tracks to the upper reaches of the Chindwin, many miles beyond. At four is the Shenam saddle and the route into the Kabaw valley. At six, the Manipur River and Log Tag Lake. At seven, the Tiddim road with the villages of Bishenpur, Potsangbam and Ninthoukhong along it. And at nine, another primitive track which wound through the Naga Hills to Silchar in Assam, 60 miles distant.

Around this perimeter fierce fighting would rage until the end of June, but the Japanese came closest to success on 10 April, when they captured Nunshigum. This isolated ridge towered 1,000 feet over the plain, dominating several track junctions and the vital airstrips upon which IV Corps relied for its survival. A counter-attack failed on 11 April and a larger attack was planned for the 13th, to be delivered by 1/17th Dogras and the Carabiniers' B Squadron, with heavy artillery and air support.

Nunshigum ridge is 7,000 yards long and its summit consists of a number of hillocks joined by hog's back cols, some only a few yards wide. The crest is covered with fairly open jungle, and the slopes with long grass and scrub. It was a hard scramble even for the infantry, but Brigadier Scoones had trained his crews in hill-climbing and was confident that the Lees were capable of tack-

1/17th Dogras and B Squadron 3rd Carabiniers commence their assault on Nunshigum. (IWM)

ling the slope.

The climb began at 10:30 and by 11:15 tanks and infantry were closing on the first of the summit features, on which shells and bombs were still bursting. This fell without undue difficulty, as did a feature in the centre of the ridge, but beyond this the way ahead to the final objective lay along a razor-backed col and it was clear that the tanks could only proceed in single file with the infantry deployed on either side. As the tanks edged forward along the col the artillery and air support programmes ended and at this moment the Japanese counter-attacked from either flank. They were held off, with great difficulty, but all of the Carabinier officers and the Dogras' two British company commanders were killed, and the attack ground to a standstill. All now hinged on the action of the senior survivors, Squadron Sergeant Major Craddock and two Dogra VCOs (Viceroy's Commissioned Officer, the approximate equivalent of the

British Warrant Officer), Subadar Ranbir Singh and Subadar Tiru Ram. The three agreed upon a plan in which the tanks with dead commanders would move off the route, following which Craddock would resume the attack with the remainder, beating in the bunkers on the final objective with close-quarter fire, and the Dogras would go in with the bayonet. The plan worked, at the second attempt, and the Dogras left not an enemy alive on the hill. The Japanese tried to recapture the feature that evening but by then it was held in strength and they were easily beaten off.

Scoones, his corps concentrated, was now able to turn his attention to the piecemeal destruction of the enemy, dug in around the edge of the plain in a wide arc. Immediately, the nature of the fighting revealed a number of factors which Mutaguchi had not allowed for in his calculations. First, there was the volume and flexibility of the British artillery fire, which he was unable to

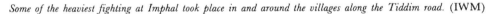

Some of the heaviest fighting at Imphal took place in and around the villages along the Tiddim road. (IWM)

match. Second, there was the unexpected ability of tanks to climb hills which might otherwise have been considered secure, ensuring the destruction of bunkers by direct gunfire. Third, the British and Indian troops enjoyed the constant and accurate close support of the RAF's Vengeance and Hurricane ground attack squadrons. In this context the ground troops marked the target with a pre-determined smoke pattern, the shape and colour of which were changed each day to prevent duplication by the Japanese. Against bunkers the Vengeance dive-bombers used delayed-action bombs which buried themselves in the structure before exploding, but against troops in the open instantaneous-fuse fragmentation bombs were most effective. Rockets were also used, each the equivalent of a medium artillery shell. For tank-busting the Hurricanes employed 40mm cannon, although targets were scarce and it was not until June that twelve were spotted and promptly destroyed. In addition to engaging battlefield targets, the RAF harassed the Japanese supply line at will, sinking river traffic on the Chindwin and strafing such transport as dared to use the roads in daylight hours. Most galling of all for Mutaguchi was the sight of an incessant stream of transport aircraft landing and taking off from IV Corps' airstrips. During the siege, the RAF evacuated 13,000 casualties and 43,000 non-combatants, brought in 12,000 reinforcements, and delivered almost 19,000 tons of supplies, ammunition and fuel. From his waterlogged trench on a hill-side above Bishenpur, Lieutenant Isobe Takuo also watched the airlift and contrasted the deteriorating condition of his men with that of the British: 'Supplies of ammunition, medicine and food were stopped by the heavy rain and the enemy air force. When we were in the jungle, we could hide from the aircraft, but we couldn't get food. We didn't have enough ammunition and were rationed to ten rounds a day. You can't expect to win

with that sort of supply situation. We had a lot of casualties from hard fighting and still more because of starvation and insufficient medicine.'

Some of the hardest fighting of the siege took place in and around the villages along the Tiddim road. These consisted of a series of compounds each containing two or three houses and possibly a well, enclosed by a cactus fence or bamboo paling standing on banks which were sometimes as much as seven feet high. Most of the houses were built on stilts and were made from bamboo matting or thin planks, but there were generally several brick or stone buildings as well. The earth banks provided excellent opportunities for digging in and the hedges severely restricted the vision of attacking tank crews. Here, too, the Japanese revealed that they could spring unpleasant surprises of their own, namely their new Model 01 47mm anti-tank gun, which was capable of penetrating the Lee at 1,000 yards. During attacks on Ninthoukhong between 20 and 25 April several tanks fell to this weapon and it was necessary to evolve tactics to meet the changed situation. The first method involved a silent night attack by infantry who secured a lodgement within the village; the tanks would enter the lodgement before first light and at dawn the methodical village-clearing drill would commence. The second method involved a daylight attack with concentrated artillery support, the tanks attacking through a curtain of smoke laid to blind the enemy anti-tank gunners.

Ninthoukhong fell on 1 May, but the Japanese became obsessed with recapturing it. Frequent counter-attacks were defeated, but at dawn on 12 June the Gurkha garrison was assaulted by a large force led by five Type 97 medium tanks. One of the garrison's two anti-tank guns was destroyed at once, but the other knocked out two of the tanks and in taking evasive action the remaining three bogged down. However, their guns remained in action and they continued to dominate the position with their fire. Reinforcements arrived, including Rifleman Ganju Lama, a Piat gunner who had already won the Military Medal for knocking out a Japanese tank some weeks earlier. The Piat, which threw a hollow-charge bomb by means of a heavy coiled spring mounted in a trough, was a difficult weapon to handle since it required considerable strength to compress the spring; normally, this could only be done by the gunner standing up and pressing with all his weight. Now, Ganju crawled forward towards the tanks, ignoring the enemy's fire, which wounded him in the leg and both arms and smashed one wrist. At thirty yards range his first bomb blasted a hole in the nearest tank and then, standing up, he recocked his weapon, inserted a fresh bomb, and crawled towards the second tank, which he also knocked out. He was about to take on the third when it was destroyed by the remaining anti-tank gun. Most men would have been more than satisfied, but Ganju had seen several survivors jump clear and they were showing signs of fight. He therefore crawled back to his comrades, obtained a supply of grenades, and returned to finish the job. His reward was the Victoria Cross.

Instead of destroying IV Corps at Imphal, Mutaguchi's divisions were bleeding to death, although Sato was successfully holding his positions at Kohima. It had been suggested that Dimapur would have been a better objective for 31st Division since its loss would also have had an effect on Stilwell's operations. This is probably true, although as events turned out it would have been over-ambitious, given the ramshackle nature of Sato's supply line.

Kohima, a small hill town, was isolated on 5 April. It was held by a tiny garrison of which the major part was formed by the 4th Queen's Own Royal West Kent Regiment. The West Kents belonged to Brigadier War-

The two Japanese Type 97 medium tanks destroyed at Ninthoukhong by Gurkha Rifleman Ganju Lama. (IWM)

Left *The Naga village at Kohima, scene of some of the bloodiest fighting of the entire campaign.* (IWM)

Right *The result of an abortive Japanese attack at Kohima.* (IWM)

ren's 161st Indian Infantry Brigade (5th Indian Division), and had only recently arrived from the Arakan. The remainder of the brigade, 1/1st Punjabis and 4/7th Rajputs, plus a handful of mountain guns, held a defensive box at Jotsoma, two miles along the Dimapur road, and this was in turn isolated on 7 April.

The battle for Kohima was fought with a savagery that has seldom been equalled. For thirteen days the little garrison beat off fanatical attacks by an enemy many times their number. Men struggled hand to hand on the surface and fought like animals in the stifling darkness as Japanese attempts to tunnel their way in were met by counter mines. Within the contracting lines some trenches were held by the dead but somehow the West Kents held on, supported by the fire of the mountain guns at Jotsoma.

Relief, however, was on the way. Lieutenant-General Montagu Stopford's XXXIII Corps, consisting of the 2nd British Division under Major-General J.M.L. Grover and the 7th Indian Division, fresh from its victories in the Arakan, had assembled in the Dimapur area and was now pushing along the road. On 14 April Grover's 5th Brigade, spearheaded by five Lees of 150 Regiment RAC manned by scratch crews, broke through to Jotsoma in the wake of a whirlwind bombardment put down by the entire divisional artillery. The Lees of A and B Squadrons 149 Regiment RAC were also on their way forward and Grover decided to await their arrival before attempting the relief of Kohima. This decision almost resulted in tragedy, for during the night of the 17th the Japanese launched a series of furious attacks which overran what remained of the position, with the exception of Garrison Hill. When morning finally broke Colonel Hugh Richards, the garrison commander, doubted whether his utterly exhausted men could survive another night, but at 08:00 Grover's artillery heralded the advance of 149 Regiment's tanks and 1/1st Punjabis; by 20 April the relief was a fact.

Stopford's next priority was to fight his way through to Imphal, but most of Kohima town and the high ground astride the main road was still held by the Japanese. Wide left and right hooks through the hills made limited progress but were halted. For the remainder of April and the first three weeks of May XXXIII Corps fought a fierce battle

of attrition as it strove to force a way through the mountain wall. The first major cracks in the Japanese defences appeared on 12 May when a number of Lees managed to work their way past Garrison Hill and destroy bunkers on Kohima Ridge; the following day a combined infantry/tank attack secured the District Commissioner's bungalow and tennis court, the scene of some of the most vicious fighting of the entire battle, and more bunkers were destroyed on the main ridge.

Sato's men were now in even worse condition than the rest of Fifteenth Army, but starving and diseased, they held on for another fortnight. In the overall context, Mutaguchi had begun to realise that he was no longer controlling the battle and he blamed his subordinates for the failure. Yamauchi and Yanagida were unfairly criticised for their lack of aggression and curtly dismissed, and the relationship between Sato and the army commander degenerated into a furious argument. Sato had repeatedly asked Mutaguchi for additional food and ammunition, without response. Instead, Mutaguchi ordered Sato to send about one-third of his division to Imphal, but the latter

declined to comply as his own losses had been disastrous. On 30 May, after a final request for supplies had been ignored, Sato informed Fifteenth Army Headquarters that he intended withdrawing. The enraged Mutaguchi promised him a court martial if he did so, to which Sato insolently replied that he would bring him down as well; for good measure, he then informed Kawabe, the army group commander, that the ability of Mutaguchi's staff fell short of that of cadets.

The survivors of Sato's division withdrew towards Chindwin, pursued by 7th Indian Division. This left the 15th Division sandwiched between the advancing XXXIII Corps and IV Corps at Imphal, and it too was forced to withdraw. On 22 June Stopford's advance guard made contact with Scoones' outposts and the siege was over. Bowing to the inevitable, the 33rd Division also headed for the Chindwin. Isobe Takuo recalls that when the retreat began, the strength of 215th Regiment amounted to a mere tenth of that with which it had embarked on the offensive, and that when it ended that figure had itself been reduced by half.

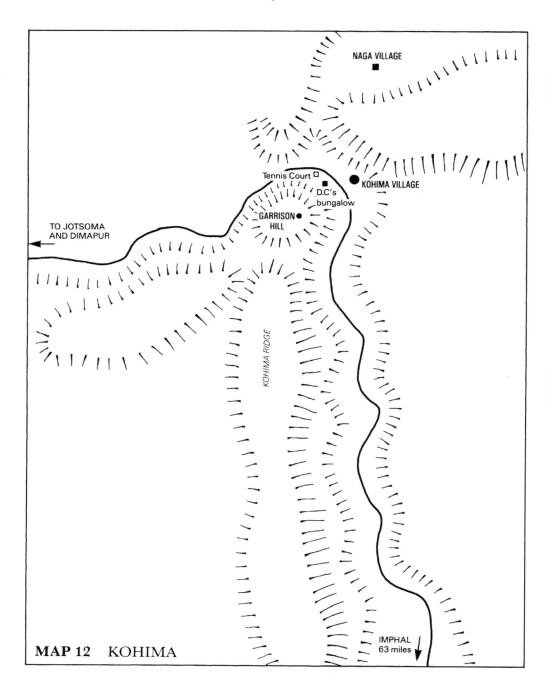

NAGA VILLAGE

Tennis Court

D.C's bungalow

KOHIMA VILLAGE

GARRISON HILL

TO JOTSOMA AND DIMAPUR

KOHIMA RIDGE

IMPHAL 63 miles

MAP 12   KOHIMA

Once a village had been captured, the tanks remained until the infantry had consolidated the position and brought up their own anti-tank guns. (IWM)

U-Go had been an unmitigated disaster. The line of the Japanese withdrawal was marked by countless bodies, abandoned tanks, guns and equipment of every kind. In field hospitals, the wounded had been shot to spare them the dishonour of falling into Allied hands. The Fifteenth Army sustained the loss of 53,000 dead, of whom perhaps half were battle casualties. Kawabe, Mutaguchi and Sato were all dismissed; the last avoided a court martial simply because it was not in the national interest at a time when Japan's fortunes were everywhere in decline. British and Indian casualties at Imphal and Kohima amounted to 17,000, but because of the efficiency of the medical services many of the wounded returned to duty later in the campaign.

In northern Burma Stilwell had renewed his advance down the Hukawng valley in February. His army consisted of two Chinese divisions, the 22nd and 38th, and an American long-range penetration group of three battalions trained on Chindit lines;

this was officially designated the 5307th Composite Unit (Provisional), but was universally known as Merrill's Marauders, a name conferred by the *Time/Life* correspondent James Shepley in honour of its commander, Brigadier-General Frank D. Merrill. Stilwell's tactics were very similar to those used by the Japanese in Malaya in that the Chinese, spearheaded by a tank group equipped with Stuarts and, later, Shermans, advanced along the main axis until checked. The Marauders would then embark on a wide flank march and establish a fortified block on the enemy's lines of communication. In this way Lieutenant-General Shinichi Tanaka's crack 18th Division, which had itself done so well in Malaya, was sharply defeated at Walumbum (3-7 March) and again at Shaduzup (28 March-1 April). After pausing until he was satisfied that Slim could contain Mutaguchi's offensive, Stilwell resumed his advance, hoping to capture Myitkyina before the onset of the monsoon. While the rest of the army advanced on

Left *Merrill's Marauders prepare to strike off the road in the Hukawng valley.* (IWM)

Below *In Northern Burma, Stilwell's Chinese–American army made extensive use of animal transport.* (IWM)

Mogaung, the Marauders, reduced by sickness and casualties to 1,400 men, were despatched with two Chinese regiments across the rugged Kumon range into the upper Irrawaddy valley. On 17 May they captured Myitkyina airfield and supplies and reinforcements were quickly flown in. The garrison of Myitkyina town, however, resisted all attempts to eject them and inflicted heavy casualties on the besiegers. Not until 3 August did the surviving remnants abandon their positions and escape across the river. A week later the Marauders, all of whom were ill and desperately weary, were disbanded. Stilwell had failed to keep his promise that the unit would be relieved

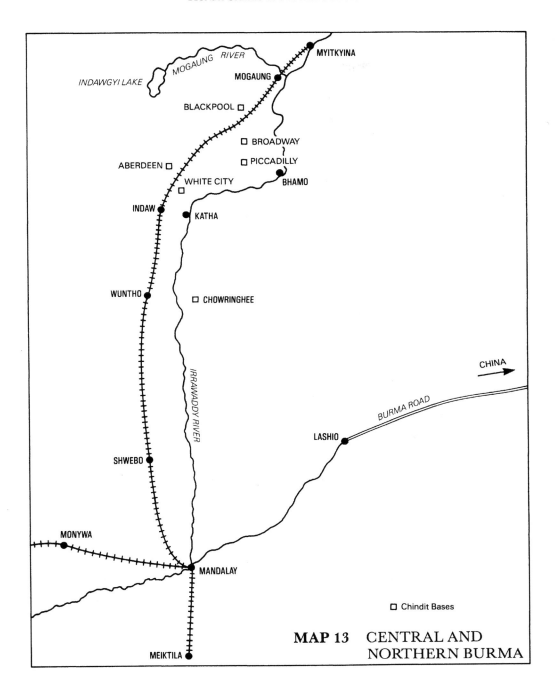

MAP 13   CENTRAL AND NORTHERN BURMA

following the capture of the airfield, and its morale was further affected by his view that genuinely sick men were not pulling their weight; he had, in fact, flogged a willing horse to death.

Further south, the Chindits had continued to raise hell, supported by the American 1st Air Commando Group under Colonel Philip Cochrane, which contained P-51A Mustang ground attack aircraft, B-25 bombers, transport, glider and light liaison aircraft. The Japanese reacted violently and eventually deployed the equivalent of two and a half divisions against the Chindits, whose total strength amounted to a single reinforced division. This was in itself an eloquent comment, since Mutaguchi had only employed three divisions against the entire Fourteenth Army, the strength of which varied between seven and a half and nine divisions. Needless to say, the troops and resources thus absorbed could have been put to excellent use at Imphal, Kohima or in the Hukawng valley.

From March until early May the principal Chindit base, codenamed White City, was located north at Indaw. A large Japanese ammunition and supply depot was captured by one unit, but it lacked sufficient demolition charges to destroy more than one dump. Instead, it blew up the medical supplies, including large quantities of quinine; the result was that Lieutenant-General Takeda's 53rd Division, which had recently arrived in the area and contained a high percentage of over-age reservists, was ravaged by malaria. The Japanese were, however, closing in on White City and Lentaigne decided to establish a new base, codenamed Blackpool and located south of Mogaung, before the monsoon set in. Blackpool was stoutly defended by 111th Brigade under Colonel John Masters, the future novelist, but by now the Chindits were extremely tired and on 25 May the base was abandoned, the sick and wounded being flown out by flying boat

from Indawgyi Lake. Despite the deteriorating condition of Lentaigne's brigades, Stilwell was bitterly opposed to their being withdrawn and simultaneously critical of their performance, and it took the personal intervention of Mountbatten to secure their relief. Even then, the 3rd West African and 14th Brigades were retained, assisting Chinese operations which culminated in the capture of Mogaung on 26 June, and were not flown out until August. The health of many was permanently damaged and some months after its return to India the force was disbanded. The Japanese verdict on the second Chindit expedition is that it was mounted at exactly the right moment in exactly the right area, to the detriment of U-Go and their operations in northern Burma.

In place of the Chindits, Stilwell was given Festing's 36th British Division, which began flying into Myitkyina during July and carried out a 100-mile advance down the railway during the monsoon. In addition to 36th Division, he now also commanded five Chinese divisions and the Mars Brigade of one Chinese and two American regiments, the latter containing a number of former Marauders. In October he mounted his final offensive. This was to result in the reopening of the Burma Road in January 1945, but by then Stilwell had been recalled to the United States at the request of Chiang Kai-shek. The cantankerous old Anglophobe had not been the easiest of allies, but neither his military ability nor his integrity had ever been in doubt. He was replaced as commander of the Chinese-American army in Burma by Lieutenant-General Dan I. Sultan.

The monsoon months had inevitably slowed the advance of Slim's Fourteenth Army following its victories at Imphal and Kohima, and they enabled General Hayotaro Kimura, the new commander of the Burma Area Army, to renew his strength and establish a defensive strategy. By

100

Right *Chindits arrive back in India. The health of many men remained permanently broken by their ordeal.* (IWM)

Below *105mm howitzer of the Chinese 38th Division in action near Mogaung.* (IWM)

November he had approximately 250,000 men at his disposal, organised in three armies: the Thirty-Third, of two divisions, under Lieutenant-General Seiza Honda, which was engaged with Sultan's forces on the northern front; the Fifteenth, of four divisions, under Lieutenant-General Shi-hachi Katamura, in central Burma and around Mandalay, still well below strength after *U-Go*; and the Twenty-Eighth, of three divisions, under Lieutenant-General Seizo Sakurai, in the Arakan and southern Burma. His plan was to allow the British to cross the Chindwin and then lure them across the Irrawaddy, where a decisive battle would be fought in the Mandalay area. During this, he reasoned, Slim's supply line would be stretched to its limits, while his own troops would be fighting close to their depots. Such a situation, he believed, would result in a reversal of the logistic nightmare which had tormented Mutaguchi, and this time it would be the British who crawled back to the Chindwin. Even had Slim not had plans of his own, Kimura's plans were wildly optimistic in that they completely

Although not intentionally designed for jungle fighting, the Lee proved ideal because of its ability to engage targets ahead and to the flank simultaneously. (IWM)

ignored the Allied capacity for air supply, an astonishing omission in view of all that had gone before.

After Imphal, Slim had reorganised his army so that Stopford's XXXIII Corps, leading the advance, contained 2nd British, 19th and 20th Indian Divisions and 254th Indian Tank Brigade, while IV Corps, now commanded by Messervy, consisted of 7th and 17th Indian Divisions, 28th East African Brigade and the all-Sherman 255th Indian Tank Brigade. He had imagined that Kimura might make a stand in the Shwebo Plain, between the Chindwin and the Irrawaddy, and in that event would have moved IV Corps in alongside XXXIII Corps. However, the fact that Japanese activity in this area was confined to rearguard actions tended to confirm that Kimura was deploying Fifteenth Army along the Irrawaddy on either side of Rangoon. Having established this, Slim was able to plan what the Japanese themselves refer to as the master stroke of the entire campaign.

It was apparent, even from the map, that supplies and reinforcements for Kimura's Irrawaddy Line had to pass through the communications centre of Meiktila. The town was, as Slim pointed out, the wrist through which lifeblood flowed into the Japanese fist clenched around Mandalay. Slash the wrist, and the nerveless fingers would open of their own accord. It was decided, therefore, that IV Corps would move in great secrecy down the Kabaw and Gangaw valleys, a journey of several hundred miles, cross the Irrawaddy at Pakoku, below its confluence with the Chindwin, and take Meiktila by *coup de main*, using a mechanized column. In the meantime, XXXIII Corps would close up to the Irrawaddy as Kimura expected, while a dummy IV Corps wireless net, operating from the Shwebo Plain, broadcast a daily

diet of routine orders and operational requirements which informed the enemy's intercept operators that Messervy's divisions were moving up to take their place in the line.

XXXIII Corps began its own crossing of the Irrawaddy on 9 January 1945 when 19th Indian Division, commanded by Major-General T. W. Rees, seized a bridgehead some 60 miles north of Mandalay; on 12 February 20th Indian Division established a second bridgehead 40 miles west of the city; and a week later Major-General Cameron Nicholson's 2nd British Division landed several miles east of 20th Division. Each crossing had been carefully timed to keep the Japanese off balance and to coincide with a specific phase of IV Corps' operations, and in each case Katamura's troops reacted with frenzied counter-attacks which simply piled up bodies around the bridgehead perimeters under the combined fire of infantry, artillery and 254th Brigade's tanks. By degrees, the British went over to the offensive, capturing nearby villages, the gar-

Above *A 3-inch mortar battery of the 19th Indian Division firing concentrations on the road to Toungoo.* (IWM)

Below *The three African divisions which served in Burma (11th East African, 81st and 82nd West African) were natural bush fighters. This infantryman is a member of the 11th East African Division.* (IWM)

Above *An anti-tank gun is used to destroy a Japanese bunker at close quarters.* (IWM)

Below *Infantry patrolling in close country on the Imphal Plain.* (IWM)

risons of which invariably fought until exterminated. After dusk, the victors would deliberately withdraw and allow fresh Japanese troops to re-enter the village in the knowledge that they would almost certainly occupy the identical positions held by their late comrades. In the morning, the British would return and, since the position of every bunker and trench was known to the tank crews and infantry alike, they would systematically slaughter the new arrivals throughout the day, before retiring again. In places, this type of action was repeated for days at a time, ruthlessly exploiting the enemy's slavish obedience to orders and sucking Fifteenth Army's reserves into a one-sided killing match.

Meanwhile, IV Corps was moving steadily south, travelling under strict radio silence along little-used buffalo tracks which traversed sparsely populated and difficult country. Progress was slow, as the route required heavy work from the corps engineers, and at times the bulldozer was the spearhead of the advance. When, during the first days of February, Messervy's advance guard reached the Irrawaddy south of Pakoku and eliminated the few local garrisons in the area selected for the crossing, the Japanese believed that the new arrivals were part of a Chindit operation and, beyond deploying troops to protect the oil installations at Chauk and Yenaungyaung, took little action.

Consequently, when 7th Indian Division crossed the river during the early hours of 14 February, it experienced little difficulty in securing and consolidating a bridgehead. Into this crossed two brigades of 17th Indian Division, the Shermans of 255th Tank Brigade and the corps' armoured car regiment, ferried on rafts.

The 25-pounder gun/howitzer, one of the outstanding weapons of World War II, was the mainstay of the Fourteenth Army's artillery. (IWM)

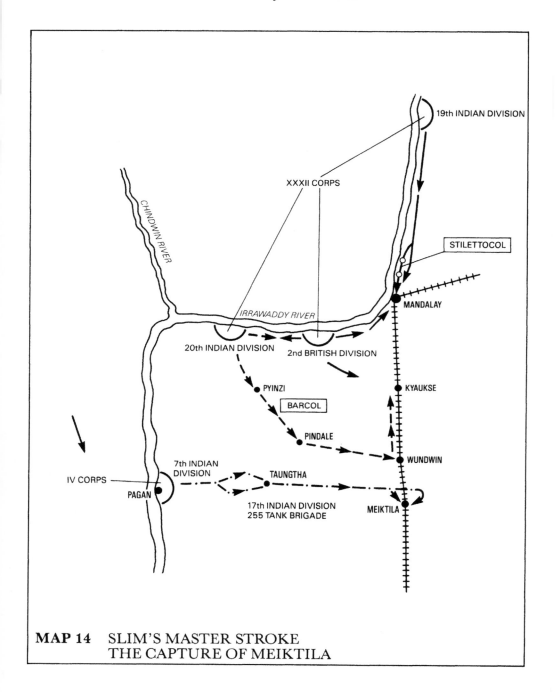

MAP 14   SLIM'S MASTER STROKE
THE CAPTURE OF MEIKTILA

By 21 February preparations for the break-out were complete and 17th Division suddenly burst through the flimsy Japanese screen, heading for Meiktila on two parallel axes. On the right, Probyn's Horse led the advance of 63rd Brigade, while to the left the Deccan Horse formed the advance guard of 48th Brigade. Ahead probed the Humber armoured cars of 16th Light Cavalry, with whom travelled the RAF's ground liaison officers, in constant communication with the cab ranks of Thunderbolts circling above, ready to bring them roaring down at the slightest sign of opposition. This part of Burma consists of semi-desert and to Messervy, flying forward in his light Auster and occasionally landing to spur on the leading elements, the sight of his tanks grinding their way through the sandy scrub must have evoked memories of his days in North Africa.

That very day the Japanese were holding a senior officers' conference at Meiktila, attended by Kimura's Chief of Staff, Lieutenant-General Tanaka, Katamura and Honda. Their discussions were interrupted by a signal which reported that a column of 200 vehicles had broken out of 7th Division's bridgehead and was heading for the town. The general opinion was that it was simply a hit-and-run raid and it was dismissed as being of no importance. In fact, in its original form the signal had put the British vehicle strength at the more accurate figure of 2,000, but this had been corrupted in transmission, with fatal consequences.

On 24 February both columns joined forces and three days later the airfield of Thabukton was captured. When 17th Division's third brigade, the 99th, was flown in from Palel on the 28th there was no further doubt in Japanese minds that Meiktila was IV Corps' objective. Major-General Kasuya, the garrison commander, halted the 168th Regiment while it was marching through on its way to the Irrawaddy Front, issued weapons to every man who could stand, including the patients at the military hospital, and gave orders that every building was to be turned into a fortress. It was too late.

Meiktila stands on a neck of land between two lakes. On 28 February 255th Tank Brigade swept round the northern approaches, swung south across the airfield to isolate the town while 48th and 63rd

Parachute-retarded fragmentation bombs proved to be the most efficient method of destroying Japanese aircraft on jungle airstrips. (IWM)

Brigades probed their way into the western suburbs. During the next three days tank and infantry teams fought their way steadily through the blazing ruins. Kasuya's 3,500-strong garrison died to a man, the last handful wading into the shallows of the South Lake to avoid capture, where they drowned or were shot down. Some Japanese concealed themselves in craters in the road, a 250-lb bomb between their knees and a stone in one hand ready to strike the exposed primer the minute a tank passed overhead; the majority were found and killed by the tanks' infantry escort.

The full horror of what had happened suddenly dawned on Kimura. The fall of Meiktila placed both the Irrawaddy and Northern Fronts in jeopardy, and the town had to be recaptured immediately. The problem was, from where could the necessary troops be drawn? Preferably not from Fifteenth Army, which was already fully committed containing XXXIII Corps, and certainly not from the Arakan, where

Christison's XV Corps had again taken the offensive and was mounting a series of amphibious landings along the coast. The only solution lay in the virtual abandonment of the Northern Front, rushing every available unit to the area, and in appointing Honda to control the operation. The 18th and 49th Divisions were directed to converge on the town and in due course links between Meiktila and IV Corps' bridgehead were severed. The 14th Tank Regiment, the only Japanese armoured formation in Burma, was also committed but was spotted making its approach march in full daylight and all but destroyed by air attack; only seven Type 97s reached the battle area, where they were used as mobile pill-boxes. Honda was further handicapped by the fact that both his divisional commanders ran their formations as private armies, making little attempt to co-operate with each other, so that the best that could be achieved was a distant blockade.

Even had the Japanese functioned effi-

Royal Garhwal Rifles attack through a smoke-screen, Arakan 1945. (IWM)

ciently, Slim's overall planning had ensured that Cowan's division at Meiktila was in no real danger. Daily, hundreds of tons of stores were delivered by air and the few casualties were evacuated. Daily, too, columns of tanks, armoured cars, guns and lorried infantry sallied forth from the defences in aggressive sweeps which inflicted heavy casualties and broke up Honda's preparations for an attack. On 17 March the 9th Brigade from 5th Indian Division was flown in to reinforce the garrison. When, three days later, Honda mounted a weak attack against the airfield it was easily beaten off and its survivors were systematically hunted down. Even if the attack had succeeded, the effort had been made too late, for as Slim had predicted, Katamura's Fifteenth Army, starved of reinforcements, ammunition and food, had suddenly collapsed like a house of cards.

Taking advantage of the situation at Meiktila, Stopford's divisions had increased the pressure around the bridgeheads and as early at 3 March 19th Indian Division had begun pushing south towards Mandalay, led by a mechanized task force consisting of 1/15th Punjabis, a squadron of Stuarts from 7th Light Cavalry, a troop of Lees belonging to 150 Regiment RAC, an artillery battery, an anti-tank troop and supporting arms, under the overall command of Lieutenant-Colonel S. G. Gardiner. This, known as Stilettocol, made excellent progress and by the morning of 7 March was engaging the defenders of Mandalay Hill. The rest of the division came up during the afternoon and the battle for the city commenced. It continued with the utmost ferocity until 20 March when the surviving Japanese, now menaced from the west by the advance of 2nd British Division, escaped through the sewers. On 19 March a mechanized column also broke out of 20th Indian Division's bridgehead to harry the disintegrating enemy army. Known as Barcol and com-

Stilettocol. Stuarts of C Squadron, 7th Light Cavalry, drive past the bullet-riddled wreckage of an ambushed jeep on the road to Mandalay. (IWM)

manded by Lieutenant-Colonel J. Barlow, this consisted of 3/4th Gurkha Rifles, two 7th Light Cavalry squadrons, a Lee squadron from 150 Regiment RAC, two armoured car squadrons from 11th Cavalry, and 18 Field Regiment with Priest 105mm self-propelled howitzers. The capture of Wundwin by Barcol cut the Mandalay-Meiktila road and forced the Japanese to withdraw well to the east of the latter.

Slim had smashed Kimura's army group, as he intended, and his next task was to deny him any chance of recovery by capturing Rangoon, 300 miles to the south. With the monsoon in the offing, such a move was a finely balanced risk, for the torrential rain would not only make the going more difficult but also inhibit the all-important air supply and ground support when flying became impossible. He calculated that the advance must maintain an average speed of ten or twelve miles per day, even allowing for opposition and demolitions. There would be no time for the development of elaborate attacks, and positions which could not be taken quickly would have to be by-passed; similarly, there would be no time for mopping up and, if necessary, large bodies of the enemy would have to be left behind. The Fourteenth Army was quickly re-organised for the final offensive; IV Corps, now consisting of 5th, 17th and 19th Divisions, supported by 255th Tank Brigade, would follow the route of the main railway line into the Sittang valley by way of Toungoo, while XXXIII Corps, with 7th and 20th Indian Divisions and 254th Tank Brigade, would advance down the Irrawaddy.

Kimura had ordered Honda to form a defence line based on Pyawbwe, which would also serve as a rallying point for stragglers trickling south from the broken Irrawaddy Front. Before this was fully formed it was outflanked to the west by a mechanized column known as Claudcol, under the personal command of Brigadier C.

E. Pert, the commander of 255th Tank Brigade. Claudcol, consisting of 6/7th Rajputs, two Sherman squadrons from Probyn's Horse, two 16th Light Cavalry armoured car squadrons, a self-propelled artillery battery and engineers, left Meiktila on 4 April and drove south, by-passing Yindaw, which was heavily defended. After capturing Yanaung and Ywadan the column cut the main road south of Pyawbwe on 9 April. The following day it turned north, shooting up transport and recently established supply dumps in the enemy's rear areas while 17th Indian Division, with tank support, launched a series of converging attacks on Honda's main positions. The Japanese, surprised by the speed with which Slim had completed his redeployment after the Mandalay—Meiktila battles and unsettled by the presence of Claudcol in the rear, were not fully prepared to resist this new offensive and by 11 April their front had been broken.

Thereafter, there was nothing to hold IV Corps, which roared south through Yamethin, Pyinmana, Toungoo and Pegu, brushing aside or smashing through such opposition as was offered, reaching Hlegu on 3 May, having covered 300 miles in three weeks. One squadron each from 7th and 16th Light Cavalry merged into two composite light squadrons to exploit the best combination of speed, mobility and firepower that light tanks and armoured cars could produce, and one of these was always well out in front, its task being to dispose of minor opposition and capture bridges before the enemy had time to blow them. Some way behind came the corps advance guard, containing two Sherman squadrons from one of 255th Tank Brigade's three regiments (116 Regiment RAC, Probyn's Horse and the Royal Deccan Horse), an infantry battalion, riding partly on the tanks and partly in lorries, a self-propelled artillery battery and an engineer troop, usually equipped with Valentine bridgelayers, which were regarded as being the most valuable vehicles in the column. Behind the advance guard travelled Brigadier Pert's brigade headquarters, including a ground-attack air liaison unit, the reserve Sherman squadron, the rest of the self-propelled artillery regiment and engineer squadron, and the second light armoured squadron, whose troops were frequently despatched to the flanks to screen villages or positions on either side of the road. Some way behind again would come the leading infantry division's own advance guard, then the division itself, followed by the rest of the corps. The momentum of the advance was actually increased towards the end by placing the spearhead on half rations and giving priority to petrol and ammunition on the air drops.

During this period, Honda and his staff narrowly escaped capture on several occasions and were reduced to the condition of fugitives. Kimura was greatly amused by his subordinate's adventures and commented that 'Honda speaks of nothing else!' Honda, who had prepared himself for ritual suicide

during one incident, doubtless found the situation less humorous.

On the more difficult Irrawaddy axis XXXIII Corps had reached Prome, but the race for Rangoon was won by an unexpected contender when 26th Indian Division from XV Corps effected an amphibious landing at the mouth of the Rangoon River on 1 May. The next day air reconnaissance revealed that the Japanese had abandoned the city and columns passed through to effect junctions with IV and XXXIII Corps.

The Burma Area Army had been completely destroyed. Between 1 January and 14 May it had lost all its armour, 430 guns and the bodies of 28,700 of its men had been physically counted; more lay where they had fallen in the jungle, and others had been buried or burned by their comrades. The Fourteenth Army sustained 13,000 casualties during this period, of whom 2,800 were killed, and about 30 of its tanks had been destroyed or seriously damaged.

Nevertheless, there were tens of thousands of Japanese troops still present in

As Slim intended, Rangoon was captured just before the monsoon broke. Burmese villagers heave a jeep through a rain-swollen chaung near the city. (IWM)

Landing craft make their way up a tortuous chaung during XV Corps' series of landings on the Arakan coast in 1945. Psy-war launches which broadcast pre-recorded invasion noises were used to mislead the Japanese as to the real landing areas. (IWM)

Burma, and most of them were as willing to fight as ever. As individuals and in groups seldom larger than battalion size, they struggled along the jungle trails with their few guns, hoping to pass through the Shan hills under cover of the monsoon and reach the safety of Thailand. The largest groups had belonged to Sakurai's Twenty-Eighth Army, and while they had not been exposed to the annihilating effects of the twin thrusts from Meiktila to Rangoon, they had further to travel from the Arakan. The progress of these thousands of starving and diseased stragglers, many naked and barefoot, was an epic of misery which only one in three survived. They were harried by regular troops, the Shan tribes were raised against them, and dacoits were armed and paid to hunt them down. Even so, they continued to fight until after their country surrendered in August, and only their own officers, now prisoners of war, could convince them that the war was over. The emergence of these tattered skeletons, still carrying their personal weapons, marked the end of the largest and most protracted jungle campaign of the Second World War.

CHAPTER 5

INSURRECTION AND CONFRONTATION

Despite all that had happened since, it was the comparatively short period of Japanese victories that made the greatest impression on minds throughout South-East Asia, for it destroyed the mystique of Western invincibility. That the Japanese had themselves been hard masters in the vast areas they occupied merely emphasised the view of the inhabitants that they no longer wished to be ruled by foreigners, and when Japan was herself defeated there was little enthusiasm for a return to the old ways.

Those colonial powers which were unwilling to accept these altered circumstances soon found themselves involved in large-scale hostilities. When the Dutch returned to the East Indies, for example, they fought a protracted war against Indonesian nationalists and, having incurred the opposition of the United States and the United Nations, were forced to grant independence in 1949. Similarly, when the French attempted to reimpose their authority in Indo-China, this resulted in a bitter and humiliating struggle, some aspects of which are discussed in the next chapter. The British, however, were more pragmatic in their approach and, wishing to part from their former subjects as friends rather than enemies, granted independence to India and Pakistan on 14 August 1947, and to Burma on 4 January 1948, it being implicitly understood that the remaining British possessions in the Far East would be granted their independence when

the time was ripe. The question of timing was extremely important since the background to these events was one of communist expansion throughout Asia, which seriously threatened Western economic and political interests.

This was particularly true of Malaya, not only because of its rich tin and rubber assets, but also because of its strategic situation. During the Second World War a clandestine British unit known as Force 136 had organized the delivery of arms and equipment to the Malayan People's Anti-Japanese Army, a guerrilla force which had been raised by members of the largely-Chinese Malayan Communist Party. By 1945 the MPAJA was 7,000 strong and the intention was that it should be used in support of the projected British invasion. When Japan surrendered before that invasion could take place the MPAJA was stood down, although some 4,000 of its members illegally retained their weapons. For a period the communists attempted to gain power by political means, but failed to make headway and by the end of 1947 had decided to employ force.

The communist plan was based on the successful guerrilla campaigns waged by Mao Tse-tung in China. During the first phase, armed groups would establish secure bases in inaccessible areas, namely the jungle which covered three-quarters of the Malayan terrain. The second phase would involve raids of increasing intensity against targets such as plantations, road and rail communications and police posts, gradually establishing a domination over the villages. Once the countryside was in their hands, the guerrillas would initiate the third and final phase, which would consist of conventional operations against the regular British forces. According to Mao, the guerrilla should swim among the population like a fish in the sea, drawing his food, supplies and recruits from friendly villages; if there were villagers who disliked the idea, they could be murdered or terrorised into providing active assistance. This philosophy was mirrored in the organization of the communist forces which, despite the fact that they called themselves the Malayan Races Liberation Army, drew their support almost exclusively from the Chinese community. In overall command was the Politburo, which formulated policy and coordinated operations, its leading military figure being Chin Peng, who had served with the MPAJA and been decorated by the British for his trouble. Below this, the most important element was the District Committee, which was responsible for supplying rations and recruits for the regular terrorist groups in its area, and for reconstituting them if they sustained serious casualties. Responsible to the District Committee were the *Min Yuen*, who lived in the villages and helped to provide food, and the *Lie Ton Ten*, who enforced discipline within the community by means of execution and torture as well as carrying out minor acts of sabotage and subversion. Notwithstanding the essentially sound structure of its organization, the MRLA suffered from a number of serious handicaps which were ultimately to prove fatal. First, it would have to function in virtual isolation, as the Royal Navy patrolled the Malayan coastline and the frontier with Thailand was closely watched. Second, outside the Chinese community it enjoyed very little support, the slightly larger Malay population being at best indifferent to its

aims. Third, apart from a few local vegetable gardens, the armed groups in the jungle depended totally on the villages for their food and if this link was cut they would cease to exist.

Following several attacks on planters, a State of Emergency was declared on 18 June 1948. Among the various provisions of the Emergency Regulations was the right to detain without trial, made necessary by communist intimidation of witnesses, the right to search property without a warrant, which produced useful intelligence and prevented the accumulation of food stocks, and the imposition of curfews on villages known to be supporting the insurgents, which imposed a degree of hardship on their inhabitants. Most important of all was the introduction of identity cards for everyone over twelve years of age. These had to be produced to obtain food and their issue not only curtailed the terrorists' freedom of movement outside the jungle but also made it extremely difficult for them to live in the villages. Together, these measures forced Chin Peng to withdraw into the jungle by the spring of 1949.

Nevertheless, the situation was far from satisfactory. Guerrilla raids were regular and frequent, resulting in 649 known civilian deaths by the end of 1949, plus 250 reported missing. The communists began mounting attacks on village police posts, but these were countered by rapid reaction forces from nearby army bases. There were now 17 British, Gurkha and Malay battalions operating in Malaya and although these killed over 1,000 insurgents, captured 600 and received the surrender of 300 more during the same period, the District Committees were performing their work well and Chin Peng's overall strength was not affected. At this period the security forces tended to react with large-scale cordon-and-search operations and the terrorists, forewarned by their supporters, managed to

British patrol 'jungle bashing' in Malaya. Incessant pressure was one of the factors which prevented the communist Malayan Races Liberation Army making headway. (IWM)

avoid major engagements by slipping out of the net. Clearly a subtler approach was required.

On 5 April 1950 Lieutenant-General Sir Harold Briggs arrived to take up the new appointment of Director of Operations and was given wide-ranging powers by the High Commissioner, Sir Henry Gurney. Briggs established greater co-operation between the armed services, the police and the civil authorities and greatly expanded the Home Guard units which provided local defence for the villages and plantations. His greatest achievement, however, was to further isolate the communists from their supporters by means of a resettlement programme. During the world recession of the 1930s many unemployed Chinese had left the towns and set up squatter camps on the edge of the jungle, where they made a living from subsistance agriculture. When the Japanese occupied Malaya still more joined these communities, until there were approximately 500,000 squatters living around the jungle fringes. Here lay Chin Peng's infrastructure and

Briggs was determined to destroy it. Over a two-year period the squatters were moved into some 400 New Villages, where they were given construction materials, the title to their new property, and a cash sum of $100. The New Villages also contained schools, clinics and shops and were protected by a perimeter fence and police post, supplemented by Home Guards. Food was escorted into the villages, but none was allowed to be taken out; in due course, the rice ration was issued cooked so that it would go bad unless eaten within 48 hours, and tinned goods were punctured as they were sold.

The New Villages proved so popular that within a year or so many of their inhabitants had begun to rebuild their homes in permanent materials. Their effect was immediate in that the larger MRLA units could not be supported in the field and had to be dispersed into smaller sub-units. This is often regarded as the turning point of the war, but the communists' reaction was violent and, although there were now 26 British, Gurkha and Malay battalions deployed against them, 1951 saw terrorist activity reach new levels of intensity. The security forces sustained the highest casualties yet and 668 civilians were murdered or vanished without trace. On 6 October the MRLA scored a major success when Sir Henry Gurney was ambushed and killed in his car. The cost of the terrorist offensive had, however, been high, for during the same period, 1,079 guerrillas were killed, 121 were captured and 201 gave themselves up.

Shortly after Gurney's death Briggs returned home and in January 1952 it was decided to combine the appointments of High Commissioner and Director of Operations under General Sir Gerald Templar. The keynote of Templar's policy was 'winning the hearts and minds of the people'. This involved convincing the population, and particularly the Chinese element, that the government had both the capacity and the will to defeat the terrorists, that it would

continue to provide protection until that end had been attained, and that economic and social benefits followed naturally once an area had been pacified. In this last respect the expansion of the Malayan economy which stemmed from the increased demand for tin and rubber during the Korean War was of considerable assistance.

In the spring of 1952 Templar's forces went over to the offensive. Of necessity, the terrorists' jungle bases were situated no more than a few hours' march from the villages on which they relied for supplies, the two being connected by comparatively few trails. By now, intelligence sources were able to pinpoint which villages were still supplying food, and ambushes were laid along the trails, steadily eroding the enemy's strength in a series of small engagements. The weapons most commonly used by the security forces in these encounters were the No 5 (shortened) version of the Lee Enfield rifle, the Bren light machine gun, the Australian

25-pounder battery deployed to provide fire support for counter-insurgency operations, Malaya 1951. (IWM)

Owen gun, the American M2 carbine, the silenced Lanchester carbine, the 9mm Browning pistol, shotguns, and the No 36 grenade. During the later stages of the war the ground-mounted Claymore mine, which discharged a fearful spread of 900 ball bearings, was widely employed. The British Sten machine carbine was intensely disliked because of its tendency to jam or discharge itself at the slightest provocation, and because it was wildly inaccurate beyond a few yards' range. Sometimes, the security forces would use the same stratagem as bird watchers entering a hide, noisily inserting a company into a suspect area and then ostentatiously withdrawing it a day or so later; the catch was, it had left behind an ambush section which would lie beside a trail for days and nights until its prey came into view, then kill in a rapid burst of fire. If they were themselves ambushed, the security forces would immediately repond with every weapon at their disposal and attempt to regain the initiative. One of the crueller facts of jungle warfare is that better results are often obtained by wounding rather than killing, since the wounded man requires the support of probably two, and sometimes four, of his comrades to carry him to safety. When British, Gurkha or Malay soldiers were hit, they could be picked up by an ambulance on the nearest road or lifted out by helicopter from a suitable clearing, but the wounded guerrilla could only drag his painful way to his base, his path marked by an easily followed trail of blood slicks. By these means, 1,155 terrorists were killed during 1952, 123 were captured and 257 surrendered; during the same period casualties among the security forces were reduced by almost half the previous year's figures, and civilian deaths dropped by one-third.

The MRLA was now under intense pressure. The recruit began to hate the system which kept him in the jungle, half-starved, for little apparent return, and the number of desertions began to rise rapidly. The government mounted a psychological warfare campaign to further weaken his resolve with offers of amnesty and reward, distributed by leaflet showers from the air or broadcast by low-flying aircraft. Many of those who gave up were only too willing to talk and this led to frequent arrests and the collapse of communist support in their villages. Some were sent back into the jungle to talk their comrades round, the security forces being diverted into other areas while these sensitive discussions took place. Others guided patrols to their former bases. The effect was to force the terrorists deeper into the jungle and compel them to further reduce the size of their active service units.

The security forces reacted to this in a number of ways. Professional trackers were recruited, including Dyaks and Ibans from Borneo. These were extremely good at their job and very dangerous opponents; the Ibans, given the chance, would gleefully use their parangs to behead their victims or remove their scalps. From March 1953, the helicopter was also used to insert troops deep into the jungle, although the size of these operations was limited by the fact that only ten troop-carrying helicopters were available. A further disadvantage was that the clearance of LZs (landing zones) was a time-consuming business involving the use of explosives and saws to fell trees, and during these early experiments in tactical deployment fully-equipped troops had to descend ropes or flexible ladders to reach the ground. Nevertheless, the increased flexibility conferred by the helicopter was yet another factor which served to lower the morale of the insurgents.

Since 1950 a new and formidable type of soldier had been serving in the jungle. In that year the Malayan Scouts were formed under the command of Lieutenant-Colonel J. M. Calvert, the former Chindit brigade commander who had reverted to his substan-

An Iban tracker interprets signs for his patrol commander. Note distinctive jungle boots. (IWM)

tive rank with the coming of peace. Calvert subjected his men to hard and realistic training in which they absorbed the accumulated jungle wisdom of the Chindits. The strength of the unit was swelled by a Special Air Service (SAS) detachment which had been bound for Korea but were no longer needed there due to changed circumstances. In 1952 the 22nd Special Air Service Regiment was formed in Malaya from the Malayan Scouts and later augmented by squadrons from Rhodesia and New Zealand.

The SAS operated in small, balanced teams of specialists deep within the jungle. Their most important functions included the killing of terrorist leaders, intelligence gathering and the protection of the aboriginal community from whom the insurgents, denied support from their own villages, would demand food. Many of the aborigines had never seen a white man before but the SAS quickly made friends with them, mainly by means of their ability to dispense medical treatment; as a matter of policy, advances by aboriginal women were

tactfully ignored since to do otherwise incurred the risk of making at least some enemies among their menfolk. Under SAS supervision jungle forts, including an airstrip capable of handling a light aircraft, were laid out, enabling the aborigines to pursue their lives in peace. The reward was not only a flow of priceless information, for in due course some aborigines took an active part in the fight against the terrorists and in the last two years of the Emergency killed more of them than the security forces using little more than their blowpipes and poison darts.

As always, the SAS matched their techniques and equipment to the task in hand and were soon able to prove that fit men, properly supported by air drops, could remain active in the jungle far longer than had been anticipated. The supply drops were made by a Dakota releasing three parachute canisters on a straight run over a river or natural clearing, timed carefully to coincide with a scheduled commercial flight to avoid attracting attention. The most spec-

An SAS team is inserted into the Malayan jungle, 1953. The 'tree-jumping' technique was perfected during the Malayan Emergency. (IWM)

tacular technique of all was the tree jump, which was faster and less obvious than insertion by helicopter. This involved the SAS team dropping directly through the jungle canopy and, quite apart from the obvious risk of serious injury, their parachutes invariably snagged among the branches, leaving the men dangling some 60 feet above the ground. Each was, however, equipped with a long rope which he secured to the parachute harness and then let himself down.

Despite government successes, the suppression of the MRLA was a long, painstaking process, but the unremitting pressure imposed by the security forces and the greatly improved intelligence situation enabled more and more terrorist bases to be identified. These were shelled by artillery, attacked from the air and even taken under naval gunfire, and although the casualties inflicted were few it was clear to the insurgents that they must move on—yet again. Even the roads, which had once offered such tempting targets, provided poor pickings since vehicles now moved in convoy pro-

tected by armoured cars and wheeled armoured personnel carriers capable of a high output of fire.

By 1955 the MRLA had had enough. In December that year Chin Peng attended a secret meeting with Tunku Abdul Rahman, the Chief Minister of the Malayan Federation, and offered to abandon his rebellion in return for formal recognition of the Communist Party. The Tunku declined and the negotiations broke down, but shortly afterwards the British Government announced that Malaya would be granted its independence by August 1957. This cut the ground from beneath the terrorists' feet since it removed their very *raison d'être*, the more so since there were now more Malayan infantry battalions in the field than any other nationality. On Independence Day, 31 August 1957, the Tunku dealt the MRLA a mortal blow when he announced a general amnesty for all terrorists. During the next year over 500 came to surrender. Incidents became fewer and fewer as more and more areas were declared safe and on 31 July 1960 the

A Cameronian patrol display flags and headgear captured in an MRLA camp; their Iban tracker also displays his parang, honed to perfection. (IWM)

State of Emergency was officially ended.

The communist insurrection had taken twelve years to defeat. It had claimed the lives of 3,000 civilians, 1,350 policemen, 128 Malayan and over 500 British and Gurkha soldiers. It is estimated that over 12,000 members of the MRLA had taken part in active operations against the security forces, and of these 6,710 are known to have been killed, while 1,290 were captured and 2,696 surrendered. Of the rest, some died of their wounds or disease, were executed on suspicion of complicity with the authorities, or simply vanished. In the last category was Chin Peng, who was said to have retired across the frontier into the jungles of Thailand. Even the most dedicated advocate of revolutionary warfare was forced to admit that the Maoist strategy was not suited to every environment, and in the MRLA's case had been severely checked during the second phase. From the British viewpoint, the war

had been ably handled at the higher level, while at the sharp end of the conflict the excellent quality of the junior leadership had proved decisive.

Two battalions of the King's African Rifles had served in Malaya and the experience gained there proved extremely useful in dealing with the Mau Mau Emergency in Kenya, as did the lessons of Malaya generally. The origins of Mau Mau are obscure, and while it certainly amounted to more than a secret society, it never became a national revolutionary movement. Most of its members belonged to the Kikuyu tribe, whose reserve lay to the north of Nairobi. The Kikuyu had a number of grievances, many of which were valid, and of these the most vexing was the question of land tenure, the tribesmen believing that they had rented land to white farmers for a period according to their custom when the latter had, in fact, bought it. Mau Mau aimed at ending the white supremacy and bound its members by means of a grisly induction ceremony and oaths which played upon their fear of the supernatural. It possessed a well developed infrastructure which kept it supplied with food, recruits, money and medical supplies but in other respects was less sophisticated than the MRLA in that it was indifferently led, untrained in revolutionary warfare, and poorly armed. The method it favoured was a reign of sheer naked terror, the majority of its victims being Kikuyu who remained loyal to the government or who refused to give active support, these men being murdered and savagely mutilated as a warning to others. In the long term this was counter-productive in that it revolted moderate African opinion and no aid for the terrorists was forthcoming from abroad. The majority of the Mau Mau gangs lived in the dense forests covering the Mount Kenya massif and the Aberdare Range, which towered respectively 17,000 and 13,000 feet above sea level. At 9,000 feet the trees ended although bamboo continued up to 11,000 feet, where it gave way to open moorland. Un-acclimatised troops found it difficult to operate at this altitude, and at nights were bitterly cold.

By September 1952 the security situation had deteriorated so sharply that the Governor of Kenya, Sir Evelyn Baring, declared a State of Emergency. Terrorist activity continued to increase, reaching its climax during the night of 26 March 1953 when Mau Mau gangs not only overran the police post at Naivasha, releasing 150 prisoners and capturing a large quantity of arms, but also hacked to death 74 people at the village of Lari, only 25 miles from Nairobi; the bodies of a further 50, mainly women and children, were never found. More troops were rushed into Kenya until the Director of Operations, General Sir George Erskine, had at his disposal up to five British infantry battalions, up to six battalions of the King's African Rifles, the all-European Kenya Regiment, an armoured car squadron and an artillery battery. The 20,000-strong Kikuyu Home Guard was also formed and, once its members had shown they could be trusted, was issued with firearms. By the middle of 1953 aggressive patrolling and cordon and search operations had begun inflicting casualties on the Mau Mau at the rate of 100 per month, but it was clear that a long haul lay ahead.

Many of the measures which had proved effective in Malaya were adopted. As the intelligence situation improved it became apparent that Mau Mau was being directed and administered from within Nairobi and in April 1954 Erskine mounted Operation Anvil, in which some 25,000 troops and police sealed off the capital, which was subjected to a methodical search, block by block. This resulted in the detention of over 16,000 suspects, many of whom were identified by hooded informers as being Mau Mau members. Anvil destroyed the link between the city and the gangs in the forest, and it also yielded much valuable information which was put to good use in the next stage of the campaign.

This was directed at clearing the Kikuyu reserve, which was designated a Special Area. In this the security forces were subject to the common law but were empowered to stop and search, to open fire if attacked or an order to halt was defied, and to open fire as the situation demanded during the hours of curfew. In the most hostile parts of the reserve New Villages were established in which the food supply could be regulated and inhabitants protected by police and Home Guard posts. The reserve was then separated from the forest by a 50-mile fire zone one mile wide; this incorporated deep ditches, wire fences and booby traps and was backed at regular intervals by fortified police posts.

When Erskine handed over to Lieutenant-General G. W. Lathbury early

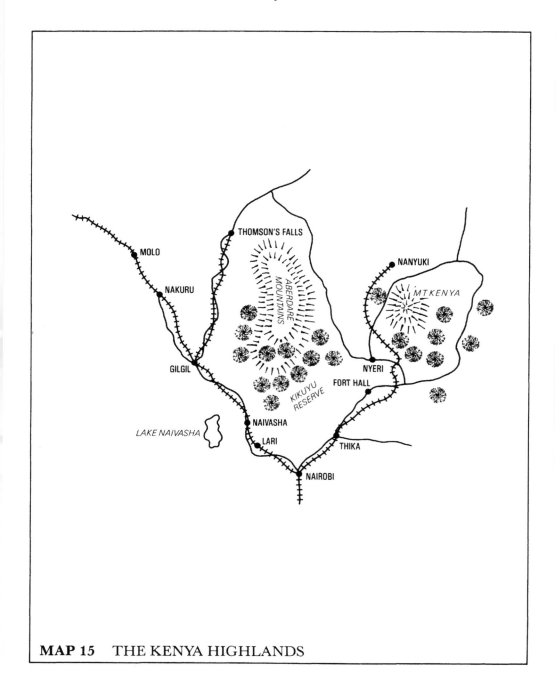

MAP 15 THE KENYA HIGHLANDS

in 1955 his measures had seriously hurt Mau Mau and forced its survivors to retire into the forest and break up their gangs into smaller units. Mount Kenya and the Aberdares were already designated Prohibited Areas in which the security forces had the right to shoot on sight and were regularly patrolled, but many gang leaders and their hard-core supporters remained at large and were difficult to root out. In addition to the daily patrolling and ambush setting, two methods were employed to solve the problem. The first involved the use of counter- or pseudo-gangs, consisting of loyal Kikuyu and led by disguised white officers and NCOs, who entered the forest posing as Mau Mau and tracked the real terrorists to their camp; many members of the counter-gangs were themselves captured Mau Mau who had been turned and were only too willing to fight against their former comrades. The second was known as the grouse drive and could involve as many as 50,000 Kikuyu civilians, including women, using their pangas to chop a way through the forest in the manner of beaters on a grouse moor, driving the terrorists steadily forward onto a stop-line manned by the security forces. The Mau Mau campaign ended with the capture of the last of the prominent gang leaders in October 1956, although the State of Emergency was maintained until January 1960 so that the detention of suspects could be prolonged.

The campaign had been won because, once again, the security forces had been able to separate the terrorist units from their supporters. Some 11,500 Mau Mau were killed in action, 2,500 were captured and 2,700 gave themselves up. The security forces lost 590 killed, of whom all but 63 were Africans. About 2,000 civilians were murdered by the terrorists, the majority being Kikuyu who remained loyal to the government.

Hardly had the Malayan and Kenyan Emergencies been brought to their success-ful conclusions than British troops found themselves engaged in a jungle war of a different kind. Ahmed Sukarno, the President of Indonesia, had long cherished the idea of creating a Greater Indonesia which would incorporate the Malay Peninsula, Singapore and the three territories of British Borneo-Sarawak, Brunei and North Borneo or Sabah. Well aware of his neighbour's ambitions, Tunku Abdul Rahman, Prime Minister of the Malayan Federation since independence, recognized that to survive his country must expand into a larger and more viable political unit which would itself absorb the Borneo territories in a Federation of Malaysia. Only the Sultan of Brunei, the population of which contained many Indonesians, expressed reservations and Sukarno promptly seized the opportunity to cause trouble. On 8 December 1962 a 4,000-strong group calling itself the North Kalimantan National Army (Kalimantan being the name of Indonesian Borneo) staged an insurrection which captured several areas of Brunei Town and a number of outlying settlements. The Sultan's request for assistance was promptly answered by the despatch of Royal Marine Commandos and Gurkhas from Singapore. By 20 December the rebels, only 1,000 of whom possessed firearms, had been defeated with the loss of 40 killed and 2,000 captured; the remainder fled into the jungle, where they were soon tracked down.

Thwarted, Sukarno declared that a state of confrontation existed between Indonesia and Malaya, the government of which he accused of pursuing a hostile policy on behalf of the former colonial power. Early in 1963 groups of Indonesian 'volunteers', bent on sabotage and subversion, began slipping across the frontier into Sarawak and Sabah, and when the Federation of Malaysia became a fact on 16 September of that year little doubt remained that a shooting war would follow.

The British, however, now had a wealth

of jungle warfare experience at their disposal and they were prepared. In December 1962 Major-General Walter Walker was appointed Commander British Forces Borneo and since then he had established the infrastructure which would enable him to meet the threat. His command was steadily expanded until, by March 1965, it consisted of 13 infantry battalions, about half of which were British and the rest Gurkha, Malaysian, Australian and New Zealand. The ground troops included a composite SAS regiment, which included British, Australian and New Zealand squadrons, the

1,550-strong Border Scouts, recruited from indigenous tribes, two battalions of the Police Field Force, and two regiments each of armoured cars, artillery and engineers. In addition, there were 80 helicopters and some 40 fixed-wing aircraft which included Javelin jet fighters capable of giving close ground support. Offshore and on the rivers, the Royal Navy maintained a fleet of coastal minesweepers and patrol craft. Walker had seen extensive service during the Malayan Emergency and set up a joint headquarters to co-ordinate the activities of the three services, the police and the intelligence-gathering agencies.

Initially, the British strategy was simply to defend the 900-mile frontier separating Sarawak and Sabah from Kalimantan. For the most part, this ran through wild mountain ranges covered in thick jungle and in places was so inaccessible that parts remained unmapped. Superficially, therefore, the task seemed impossible with the limited resources available. On the other hand, the terrain was just as difficult for the enemy, whose approach could only be made along

established trails which could be watched, or in longboats on the rivers which formed the principal highways of the area.

With this in mind, it was essential to win the goodwill of the border tribes before the Indonesians attacked. This was achieved by the SAS teams, many of whom had become fluent in Malay during the Emergency. They lived in the settlement long-houses for weeks at a time, paying their way, providing medical care and making friends, so that the Indonesians were barely able to make a move along the frontier without it being reported. The border itself was defended by a series of fortified bases which were sited to cover the likely crossing places, a mile or so inside Malaysian territory. These were usually held in company or platoon strength and were surrounded by a perimeter of barbed wire and sharpened bamboo stakes known as *panjis*, capable of disembowelling a man. The growth around these forts was cleared to give a good field of fire and the probable avenues of attack covered by Claymore mines. Within, bunkers and trenches protected the garrison,

Italian-designed 105mm pack howitzer in action at a British base in Sarawak. The same weapon was used by Australian and New Zealand troops in South Vietnam. (IWM)

and emplacements were constructed for one or more 81 mm mortars or light 105 mm howitzers, which were lifted into position by Belvedere helicopters; whenever possible, forts were situated so that they could provide each other with mutual artillery support. From these bases extensive patrols were carried out, sometimes lasting for several days. The use of sensitive seismic detectors, capable of registering footfalls, also provided the defenders with remote advanced listening posts. The bulk of Walker's troops, however, were stationed in camps further back and were employed as a rapid reaction force which was capable of deployment by helicopter into a threatened sector. As a result of these measures, the Indonesians made very little headway. In September 1963 a 200-strong group of 'volunteers' overran a small outpost at Long Jawai, but 1/2nd Gurkha Rifles were lifted by helicopter into ambush positions along the raiders' line of retreat and very few of the party succeeded in regaining the frontier. Three months later a large group which attacked the village of Kalabatan was similarly dealt with in a

month-long operation from which only six of the invaders escaped.

By March 1964 it was apparent that Sukarno had replaced his 'volunteers' with regular army units and the pressure on the British and Malaysian forces began to increase. By September it was also clear that the Indonesians had established bases of their own across the frontier. The British government was at first reluctant to permit cross-border strikes to eliminate these, but relented when Sukarno began inserting parties onto the Malayan mainland, all of whom were quickly rounded up. Forays into Indonesian territory were generally made in company strength against objectives which had been carefully reconnoitred by the SAS, with artillery but normally without air support. Initially, such raids were sanctioned to a depth of 2,000 metres, but this was later raised to 10,000 metres and, in appropriate circumstances, to 20,000 metres. As a result, the Indonesians were forced to abandon their forward bases. This offensive policy was maintained by Major-General George Lea when he took over from General Walker

Above *Contour-flying heli-copters transport troops to a British base camp near the Indonesian border.* (IWM)

Right *An RAF Westland Whirlwind helicopter co-operates with Ferret scout cars of 4 RTR during an area reconnaissance patrol.* (IWM)

in 1965. Despite being small in scale and cloaked in secrecy these actions were fiercely contested and proved the efficiency of the light anti-tank rocket launcher when employed against machine gun posts and buildings.

In addition to guarding the frontier, the security forces also had to maintain watch on the 24,000-strong Clandestine Communist Organisation (CCO), which drew most of its strength from the urban Chinese population. The CCO had a potential for sabotage and subversion which Sukarno would gladly have mobilised, although on its own it was too weak and scattered to have achieved much. Ironically, it was Sukarno's encouragement of the Indonesian Communist Party which, together with social and economic pressures, led to his downfall. In March 1966 he was stripped of his powers by the Army and five months later his successor ended the state of confrontation with Malaysia.

During the undeclared war with Indonesia casualties sustained by the Commonwealth forces amounted to 114 killed and 181 wounded. The Indonesian loss almost certainly exceeded 700 killed, plus 771 captured. Notwithstanding the limited nature of the operations involved, the campaign remains one of the most efficient ever conducted in a jungle environment.

CHAPTER 6

THE ROAD TO DIEN BIEN PHU

Modern Vietnam was formed from the three ancient kingdoms of Tonkin in the north, Annam in the centre and Cochin China in the south. Its coastline is over 1,200 miles in length, stretching from China to the tip of the Indochina peninsula, and its equally long land frontier borders China, Laos and Cambodia, now known as Kampuchea. In the north, Vietnam has a maximum width of 300 miles, but for most of its length is barely 130 miles wide. The country contains five distinct geographical regions: the Northern Mountains, running along the Chinese and Laotian borders, with peaks up to 8,000 feet in height; the Northern Plains, including the Red River Delta and Coastal Lowlands; the Central Highlands, consisting of rugged, mountainous country which is the home of the indigenous peoples unrelated to the Vietnamese, with high, rolling plateaux further south; the Coastal Lowlands, a narrow strip stretching from the Northern Plains to the Mekong River Delta, subdivided by spurs from the Central Highlands; and the Southern Plains, which include the Mekong River Delta and the surrounding fertile plain, intersected by a network of rivers and irrigation canals. Each of these regions contains areas of jungle, swamp, grassland and cultivation, including rice paddies. From May until early October the summer monsoon produces hot, cloudy, humid conditions and a monthly rainfall of up to 100 inches, concentrated mainly on the high ground.

During the winter monsoon, which lasts from early November until mid-March, it is the coastal areas which receive the heaviest precipitation.

The French had arrived in Cochin China in 1862 and by 1883 had extended their control over the whole of Vietnam. France's defeat in 1940, followed by the Japanese occupation of Vietnam and the reverses sustained by the British, Americans and Dutch during the next two years, all combined to encourage the aspirations of nationalist groups within the country. Of these the most important was the *Viet Nam Doc Lap Dong Minh Hoi* (Vietnam Independence League), more commonly shortened to Viet Minh, formed in May 1941 by the Indochinese Communist Party as a broadly based national liberation front. At the head of this was a veteran communist who had lived for seven years in Paris, prior to completing his training in Moscow; he had been born in 1890 as Nguyen Tat Thanh, although he used many aliases, but the name by which he became universally known was that of Ho Chi Minh, which means He Who Enlightens. Ho entrusted the task of creating the Viet Minh's military wing to one of his closest associates, Vo Nguyen Giap, a history teacher noted for his organizational ability and implacable will.

At this stage the Western Allies believed that the Viet Minh were solely an anti-Japanese resistance group and willingly instituted a supply of arms, although for Giap the real enemy remained the French, who still nominally ruled the country. He continued to build up his strength and in January 1945 captured several gendarmerie posts in Cao Bang province. Normally the Japanese were prepared to leave internal security matters to the French, but they were now becoming seriously concerned by the possibility of Allied landings in Vietnam, following which they believed that the Vichy authorities would change sides. On 9 March, only three days before the French were due to commence operations against Giap, the Viet Minh staged a coup throughout Indochina. The majority of French troops were disarmed and interned, those that offered resistance being massacred; the Foreign Legion's 5th Infantry Regiment fought its way through numerous ambushes to the Chinese frontier, losing two-thirds of its strength in the process.

The Allies had agreed at the Potsdam conference that when Japan surrendered Vietnam would be occupied by the British as far north as the 16th Parallel and by the Nationalist Chinese beyond that. In the south, the British handed control back to the French, who quickly re-asserted their authority, but in the north the Chinese refused to permit the return of the colonial power for a while. During this period the Viet Minh occupied Hanoi and on 2 September Ho Chi Minh proclaimed the independence of the Democratic Republic of Vietnam. Simultaneously, Giap's troops acquired large quantities of surrendered Japanese arms, augmented by supplies from China. On 6 March 1946 the Paris government reached an agreement with the Viet Minh which recognized the Democratic Republic, but only within the French Union. This fell far short of Ho's ideal of independence, but he accepted it as he believed that the French could soon be defeated, whereas it would be difficult to eject the Chinese if

LVT-4 Alligators were used as APCs by the French Groupements Autonommes *in the Red River and Mekong Deltas.* (ECP Armées)

they remained much longer. French troops began landing immediately, the Chinese withdrew within their frontier, and the Japanese were shipped home.

For a while the French and the Viet Minh co-existed uneasily in Hanoi. In November a clash in Haiphong led to the port being bombarded by French warships, with heavy civilian loss of life. Fighting became general and by the end of the year the Viet Minh had been forced out of Hanoi itself. The scene was now set for a major armed conflict, which Ho and Giap intended waging both at the political and military levels.

The public image of the avuncular Ho might have been that of the benevolent oriental sage, but this concealed a ruthlessness comparable to Stalin's. Most of the population lived in villages and his first task was to separate them from the French administration. This was achieved partly by indoctrination and partly by murdering village headmen, local officials, schoolmasters, and indeed anyone who owed their position to the establishment, unless they provided active support. Likewise, the leading figures of alternate political parties were efficiently eliminated by assassination squads. Bomb and other outrages were stage-managed in the towns and cities to demonstrate French impotence. Concurrently, Giap continued to build up the already considerable strength of the Viet Minh's military wing. The country was divided into regions, each of which was controlled by a committee responsible to Ho and his revolutionary government for raising and training the local guerrilla forces. These included the village militias, which had a low military potential but were useful in providing support for regular units in the field, for intelligence gathering and as a source of labour, and the regional troops, who were better armed and trained and were used to mounting raids on French garrisons and convoys, taking full advantage of their

intimate knowledge of the countryside. Neither wore uniform, and since both lived in the villages they were impossible to distinguish from the local population. Giap's regular troops, known as the Chuc Luc, were well armed and trained and distinguished by their uniform of black pyjamas and cork helmets. Their role was to defeat the French in open warfare once the latter had been decisively weakened during the guerrilla phase of the campaign. In 1950, thanks to a flow of arms from China, Giap was able to form his Chuc Luc battalions into divisions, each consisting of three two- or three-battalion infantry regiments plus heavy mortar, anti-aircraft and engineer units. As the war progressed, it became possible to add field artillery to the divisional establishment. The entire Viet Minh military machine was constructed to implement the Maoist theory of revolutionary warfare and therefore possessed great flexibility; it was also difficult to damage, since it enjoyed widespread if not universal popular support, usually retained the initiative, fought only when it knew it could win and avoided contact with superior forces.

The French, on the other hand, fielded a conventional army which included armour, artillery and air support, but was never large enough to suppress a major uprising of this kind. The army also suffered from a number of serious handicaps. Its senior officers were understandably anxious to restore the honour of French arms after the traumas of the Second World War, but bitterness still existed between those who had fought with De Gaulle and those who had remained loyal to the Vichy government. They were, too, handicapped by decisions made in France, where political expediency was a more valuable coinage than military necessity. For example, conscripts could not be employed in Vietnam, and this restricted the purely French element of the army to the number of units

131

The smaller M29 Weasel, known as the Crabe in the French Army, served as a weapons carrier in the Groupement Autonommes. This example carries the insignia of the Foreign Legion's 1st Cavalry Regiment. (ECP Armées)

which could be formed from regular volunteers. Again, during a period of disastrous defeats in 1950, the government of the day actually *reduced* the size of the army in Indochina by 9,000 men. Worst of all was the fact that the French Army had no recent experience of jungle warfare, let alone counter-insurgency operations on this scale. On average, it deployed 150,000 men in Indochina (50,000 French, 20,000 Foreign Legion, 30,000 African and 50,000 Vietnamese) but, like the British in 1941-42, it was mentally round-bound, and of these more than half were tied down in static garrisons or employed on convoy escort duties. For offensive operations the French formed seven mobile groups equipped with tanks, armoured cars, half-track armoured personnel carriers and Weasel and Alligator amphibians, depending on the nature of the going; there were also eight parachute battalions which were regarded as the cutting edge of the French response. Air support was provided by Marauder and Privateer bombers and Bearcat and Corsair fighters. Transport aircraft were in critically short supply, the tri-motor Junkers Ju-52 being used until replaced by Dakotas and Flying Boxcars in 1952. Latterly a number of H-19B helicopters also became available.

During the first phase of the war Giap concentrated on guerrilla activity. The French held the urban areas, the fortified posts between, and—by day—the roads. At night, the countryside became the province of the Viet Minh, particularly in the north, where their support was strongest. The isolated garrisons and the convoys which supplied them made excellent targets for attacks and ambushes, and by the time a relief column had arrived the raiders had merged back into the landscape. Little intelligence was forthcoming from a population which had either been intimidated into silence or was sullenly hostile. The Viet Minh, on the other hand, were aware of every move the French planned. Thus, when cordon-and-search operations were mounted, often with

troops drawn from garrisons elsewhere, the local guerrilla units received plenty of warning and were able to slip clear of the net to harass the areas of weakness created by the re-deployment. Most engagements were comparatively minor, with the Viet Minh breaking contact if faced with serious opposition. How long the stalemate might have lasted is a matter for speculation, but its end was apparent in 1949 when Chiang Kai-shek was finally defeated by the Chinese communists. This ensured that the Viet Minh would receive all the arms and support it needed from across the frontier.

Giap's first task was to clear his lines of communication by capturing Cao Bang ridge, which lay adjacent to the border. Along the ridge ran the Route Coloniale 4, linking the French garrisons of Long Son, That Khe, Dong Khe and Cao Bang. The French High Command had, in fact, already decided that these isolated outposts were untenable and had evolved a plan for their abandonment. This involved the garrison of Cao Bang, consisting of 1,000 men of the Legion's 3rd Infantry Regiment and 600

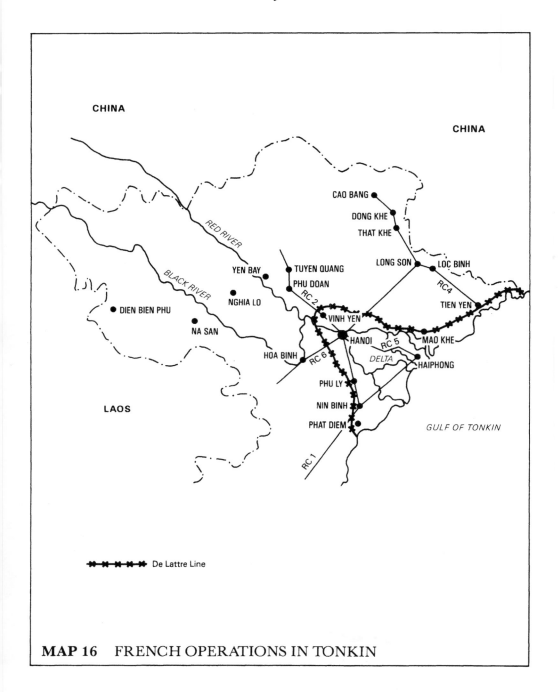

MAP 16 FRENCH OPERATIONS IN TONKIN

Moroccans, accompanied by the civilian population which was hostile to the Viet Minh, marching the 15 miles to Dong Khe, where it would be met and escorted to safety by a force composed of three Moroccan battalions and the Legion's 1st Parachute Battalion. Giap, informed of the details of the plan by his excellent intelligence service, realized that an opportunity existed to inflict a major defeat on the French. The small fort at Dong Khe, held by two companies of the Legion's 3rd Infantry Regiment, was shelled throughout 16 September 1950 and the following day was assaulted by two regular Viet Minh regiments. True to their traditions, the Legionaires fought to the bitter end, a handful of survivors escaping into the jungle when the defences were finally overrun on 18 September. Meanwhile, the 1st Parachute Battalion had jumped at That Khe and its probes northward had encountered stiff opposition. Intelligence reports suggested the presence of at least 15 Viet Minh battalions and supporting artillery, although the truth was that Giap had deployed no less than 30 battalions for the operation.

Nevertheless, it was decided to continue with the evacuation of Cao Bang. The relief column, known as Group Bayard, began assembling at That Khe on 1 October but the following day was halted south of Dong Khe and heavily counter-attacked. The Cao Bang garrison, commanded by Colonel Charton, left during the night of 2/3 October but made very little progress before it, too, was brought to a standstill. Clearly, the original plan was unworkable and it was agreed that both columns would leave the Route Coloniale 4 and effect a junction on jungle tracks to the west. What followed bore a horrible similarity to the massacre of Varus's legions in AD 9. After four days of close-quarter fire-fights, food, water and ammunition began to run out. Stragglers were dragged off the trails and hacked to death, and in one incident men were forced to escape down lianas overhanging a sheer cliff, some with wounded clinging to their backs. The remnants of the Bayard group eventually reached Charton's column on 7 October to find it had been equally mauled. Abandoning its wounded, the combined

An M8 75mm Howitzer Motor Carriage of the 1er Chasseurs à Cheval *and supporting infantry on patrol near the Black River in December 1951.* (ECPA)

force split into small groups and attempted to reach That Khe, where the 3rd Colonial Para-Commando Battalion was dropped on 8 October. Very few got through before the post was abandoned two days later, and the Para-Commando battalion was itself ambushed and destroyed as it withdrew. The disaster on Route Coloniale 4 severely jolted French morale, both in the field and in France itself. Giap had inflicted over 6,000 casualties and captured enough equipment to equip a division, and in the immediate aftermath Long Son was evacuated, as were garrisons as far distant as Hoa Binh.

The appointment of General Jean de Lattre de Tassigny as Governor-General and Commander-in-Chief in December went some way to restoring French confidence. De Lattre recognized that for the moment he would have to remain on the defensive and concentrated his forces in the areas of greatest strategic importance, including the Mekong Delta, Tourane and Hue in Annam, and the Red River Delta around Hanoi, where he constructed a chain of fortified

positions which became known as the De Lattre Line. Supplies of American arms and equipment also began reaching Indochina, for since the outbreak of hostilities in Korea the United States had come to regard the campaign as part of the struggle to contain the spread of communism rather than as a colonial war.

For his part, Giap over-estimated the significance of his recent victory in northern Tonkin and demonstrated one of the weaker aspects of his generalship. Believing that the French had been fatally weakened, he decided to resort to conventional warfare and in January 1951 launched an offensive against de Lattre's positions in the Red River Delta. The first assault, lasting from 13-17 January, involved two Viet Minh divisions and was directed at Vinh Yen, some 40 miles north-west of Hanoi. After some early successes the attackers were halted by reinforcements airlifted from other areas of the delta, and then decimated by concentrated artillery fire and air strikes which made extensive use of napalm. Having failed at

Outside Tonkin, support for the Viet Minh was uneven. Here a French-equipped Vietnamese unit trains for local defence. (SIRPA/ECPA)

French infantry probe their way cautiously into a village. Many Viet Minh supporters were farmers by day and guerrillas only by night, but the danger of ambush was always present. (SIRPA/ECPA)

Vinh Yen, Giap attempted to capture Mao Khe, north of Haiphong, on 23 March, using three divisions. Again, an early penetration was made but on 26 March French warships on the Do Bac River fired a devastating bombardment into the massed ranks of the 316th Division while it was forming up for the final assault. The subsequent pause allowed the garrison to be reinforced by air and when the attack was renewed over the next two days it failed to penetrate the curtain of high explosive and napalm. At the end of May Giap tried again, launching three divisions against Phu Ly,

M8 75mm HMC and M8 Greyhound armoured car engage a distant enemy with their heavy machine guns. (ECPA)

Ninh Binh and Phat Diem on the Day River while a fourth division attempted, unsuccessfully, to tie down French reserves with diversionary attacks elsewhere. The fighting lasted until 19 June and once more ended in a bloody repulse for the Viet Minh. Having incurred not less than 20,000 casualties during the abortive offensive against the De Lattre Line, Giap was forced to concede that his foray into conventional warfare had been premature, and reverted to guerrilla activity. One reason why the Viet Minh sustained such heavy losses was that, like the Chinese in Korea, they employed the 'human wave' method of attack, hoping to swamp the defences under sheer weight of numbers; given the volume of firepower available to the French, this was a recipe for disaster.

Having said this, it is also necessary to emphasise that in both the war against the French and later the Americans in Vietnam the communist forces were prepared for and willing to accept an abnormally high rate of attrition to secure their objectives.

In October Giap tried to capture Nghia Lo, lying on the ridge between the Red and Black Rivers, but paratroops were dropped to reinforce the garrison and, once again, the Viet Minh were forced to retire. On 14 November the French went over to the offensive. Three parachute battalions dropped at Hoa Binh, 50 miles west of Hanoi, which had become an important Viet Minh supply depot. The town was captured without difficulty and mobile groups established a corridor to it along the Black River and Route

Signallers at work in a temporary command post. (SIRPA/ECPA)

Above *An M5 Stuart grinds its way past African colonial infantry during the monsoon season.* (ECPA)

Right *Ambush! A remarkable photograph showing an immediate reaction. The smoking craters and the position of the casualty suggest that the unit has come under mortar fire.* (SIRPA/ECPA)

Coloniale 6. Giap recognized that by establishing new posts to defend these routes the French had over-extended themselves and, resisting the temptation to recapture Hoa Binh by direct assault, he preyed incessantly on their lines of communication, inflicting serious casualties in the process. At this point de Lattre, suffering from terminal cancer, returned to France, where he died shortly after. His successor, General Raoul Salan, realized that Hoa Binh was a wasting asset and decided to cut his losses, but such was Giap's grip on Route Coloniale 6 that it took twelve battalions, supported by three artillery groups and air strikes, from 18 until 29 January 1952 to re-open the road. Even then, when the garrisons of Hoa Binh and the intermediate posts were withdrawn

between 22 and 24 February, they were forced to fight their way through one ambush after another.

On 17 October the Viet Minh 308th Division finally captured Nghia Lo after a bitter struggle, although Giap's further progress west towards the Laotian frontier was checked by the tenacious defence of Na San. Meanwhile, Salan was preparing a counterstroke which he hoped would force the Viet Minh to fight a decisive battle on his terms. The heart of the problem lay in concentrating sufficient French troops, and to this end a Vietnamese National Army had been formed with a view to releasing French units for offensive operations. Vietnamese serving with French colonial units invariably fought well, but the Vietnamese National Army suffered from a shortage of suitable officers and its men tended to desert if required to serve outside their own provinces. Neverthe-

less, by the end of October a 30,000-strong French task force had been assembled and Salan intended using this to strike out from the Delta deep into Viet Minh territory, forcing Giap to abandon his own plans and defend his supply bases.

The operation, codenamed Lorraine, began on 29 October when two mobile groups began advancing up Route Coloniale 2 towards Phu Doan, 100 miles north-east of Hanoi. The advance was contested only by the Viet Minh's regional troops and steady progress was made. On 9 November the Legion's 1st and 2nd Parachute Battalions and the 3rd Colonial Parachute Battalion were dropped on Phu Doan itself, where large quantities of arms and supplies were discovered and destroyed. Once contact with the ground forces had been established strong patrols were pushed out westwards to Yen Bay and northwards to Tuyen Quang in

Covered by its light machine gun section, an infantry squad changes position. The man closest to the camera is armed with a rifle grenade. (SIRPA/ECPA)

Above *The quality of the purely Vietnamese element of the French army improved steadily towards the end of the war, but there was always a shortage of officers. A Vietnamese unit looks for any intelligence to be found on regular Viet Minh casualties.* (SIRPA/ECPA)

Below *Parachute troops formed the cutting edge of the French response to the Viet Minh.* (SIRPA/ECPA)

Above *Paratroop commanders confer during a pause in operations. The machine pistols, grenades clipped to webbing, machete, field dressings strapped to helmets and the vigilant sentry in the background all indicate a thoroughly experienced unit.* (SIRPA/ECPA)

Left *White scout cars head a convoy bound for a French outpost.* (ECPA)

anticipation of Giap's counter-attack. Giap, however, refused to be drawn and directed his response at the French lines of communication, just as he had at Hoa Binh the previous year. The *Armee de l'Air*, already stretched to the limit supplying distant garrisons, could not support the additional burden and on 14 November Salan, accepting that his gamble had failed, began to withdraw. The Viet Minh staged a number of successful ambushes, notably in the Chan Muong gorge on 17 November, inflicting heavy casualties which included 1,200 killed, before the French regained the safety of the

Closely escorted by paratroopers, a Sherman moves through ideal ambush terrain. The survival of both armour and infantry depended upon constant vigilance. (SIRPA/ECPA)

Delta defences. The Legion's 1st and 2nd Parachute Battalions were immediately flown up to Na San, where their arrival proved decisive. On 2 December the Viet Minh 312th Division abandoned its assaults on the post, leaving 600 bodies entangled in the wire, plus a litter of recently arrived Chinese weapons.

Giap next sought to further over-extend the French by supporting the growing nationalist movements in neighbouring Laos and Cambodia. In April 1953 two Viet Minh divisions invaded Laos and Salan was only able to halt their advance on the capital, Luang Prabang, by rapidly transferring much of his small reserve to the Plain of Jars.

The following month Salan was succeeded by General Henri Navarre as Commander-in-Chief. Having assessed the situation at first hand, Navarre believed that he could hold the Delta and Cochin China and, having built up the Vietnamese National Army for defensive tasks, could take the offensive in 1955, provided he received adequate reinforcements. He went to France to plead his case personally but the government of the day, sensitive to the anti-war sentiment generated by the French Communist Party, which made political capital of the casualties already incurred and the expense involved, declined to make more than ten battalions available. This was well below what was considered necessary to achieve a decisive result yet, perversely, the government also refused to open negotiations with Ho Chi Minh. Navarre, perforce, must continue to make bricks without straw.

Nevertheless, in July he mounted a brilliantly successful raid against the Viet Minh supply dumps at Lang Son, lying deep within territory which Giap had long regarded as secure. The operation, codenamed Hirondelle, was planned in secret by its commander, General Gilles, and the troops involved were not briefed until the last possible moment. At 08:10 on 17 July the 6th and 8th Colonial Parachute Battalions jumped over Lang Son, routing the surprised local troops. In caves surrounding the town were large quantities of arms, ammunition and supplies, which were systematically des-

Wounded paras assembled at a casualty clearing post. (SIRPA/ECPA)

troyed. In the meantime the 5th Mobile Group had left the nearest point on the De Lattre Line, Tien Yen, and was pushing forward along the Route Coloniale 4 while the Legion's 2nd Parachute Battalion dropped at the intermediate point of Loc Binh, securing a crossing over the Song Ky River. Having completed their work of destruction, the Colonial paratroops embarked on a forced march in intense heat, accompanied by over 300 civilians who no longer wished to remain under communist rule, reaching Loc Binh at 23:00 on 18 July. The Legion then formed the rearguard until contact was established with the 5th Mobile Group at Dinh Lap, where transport was waiting, and the whole force withdrew through Tien Yen. The cost of the operation was one man killed in action, one missing, three dead from heatstroke and 21 wounded, who were evacuated by helicopter.

The following month Navarre increased the size of his reserve by withdrawing the outlying garrisons, including that of Na San, although a drive through the Coastal Lowlands failed to produce significant results because the Viet Minh avoided contact whenever possible. However, the success of French air reinforcement and air supply operations thus far, coupled with Giap's incursion into Laos, seemed to present Navarre with the opportunity for decisive action which his predecessors had long sought. He decided to establish a major fortified base at Dien Bien Phu, lying astride the Viet Minh's line of communications with Laos, knowing that Giap would have to react to the threat. The base itself would be supplied by air, would contain adequate artillery to break up enemy attacks, and the *Armee de l'Air* would harass the Viet Minh as they attempted to concentrate. In the end, it was anticipated that Giap's forces would be so written down that their offensive potential would be effectively destroyed. The choice of Dien Bien Phu seemed ideal in every respect, since it lay in the territory of the T'ai people, who were well disposed towards the French, and already contained an old airstrip. It was situated in a valley some ten miles long by four miles wide, bisected by the Nam Yum River, and contained a number of small hills which could easily be turned into strongpoints. Finally, the historical precedents of the Admin Box, Imphal and the Chindit bases in Burma, and indeed the defence of Na San the previous year, all seemed to indicate that Navarre's strategy was the correct one.

Unfortunately, the planners omitted certain critical factors from their calculations. Dien Bien Phu lay so far from the Delta that, unlike the Admin Box and Imphal, there was no prospect of its being relieved by land forces. The nearest air base was 140 miles distant, but the Chinese frontier was only 80 miles away. The valley itself was ringed by heavily forested hills up to 1,800 feet in height, and during the monsoon season was prone to flooding and thick mists and rain which made flying extremely difficult. Worse still, the possibility that the Viet Minh might possess adequate artillery and an anti-aircraft capability was largely discounted.

Operation Castor, the capture of Dien Bien Phu by the French, began on 20 November. Six parachute battalions, plus supporting heavy weapons and an engineer company, dropped on the old airstrip, driving off two Viet Minh companies which were training in the area. By 24 November the airstrip was in commission and transport aircraft began arriving in a steady stream to unload men and equipment. The low hills on the plain were turned into a loose ring of fortifications allegedly named after the former mistresses of the garrison commander, General Christian de Castries; these were surrounded by wire entanglements and minefields and contained bunkers roofed with layers of logs and earth. In the centre of

the position was the village, which housed de Castries' command bunker and an underground hospital. Strongpoint *Huguette* lay to the west, *Claudine* to the south, *Eliane* to the east and *Dominique* to the north-east; an outer ring of similarly constructed but isolated strongpoints was named *Anne-Marie* (one mile to the north-west), *Gabrielle* (two miles to the north), *Beatrice* (one mile to the north-east) and *Isabelle* (four miles to the south, covering a newly constructed auxiliary airstrip). Unfortunately, the French artillery commander, Colonel Charles Piroth, believed that the enemy would be unable to deploy heavy weapons and dug in his guns only to the extent that they were protected against mortar attack; he was to take his own life when the awful truth became apparent, but he was not alone in holding this opinion.

By the beginning of March 1954 there were over 10,000 men at Dien Bien Phu, including four Legion battalions, Vietnamese, North Africans and T'ais. The garrison possessed four 155mm guns, twenty-four 105mm guns, a number of 20mm and 40mm anti-aircraft guns for use in the ground role, ten M24 Chaffee light tanks which had been air-landed in parts and assembled *in situ*, and a small fighter-bomber squadron which operated from the airstrip. Early in the New Year patrol clashes in the surrounding hills had confirmed that the Viet Minh were arriving in strength and from 31 January sporadic artillery fire began erupting around the strongpoints and command bunkers. It was apparent that Giap was reacting just as the French High Command hoped he would and his attack was keenly awaited.

Had the French been aware of the nature of Giap's reaction they would have been less sanguine, for the whole Viet Minh regular army was now deployed against them. By degrees, the 304th Division closed in from the south, the 316th and 312th Divisions from the east and north, and the 308th

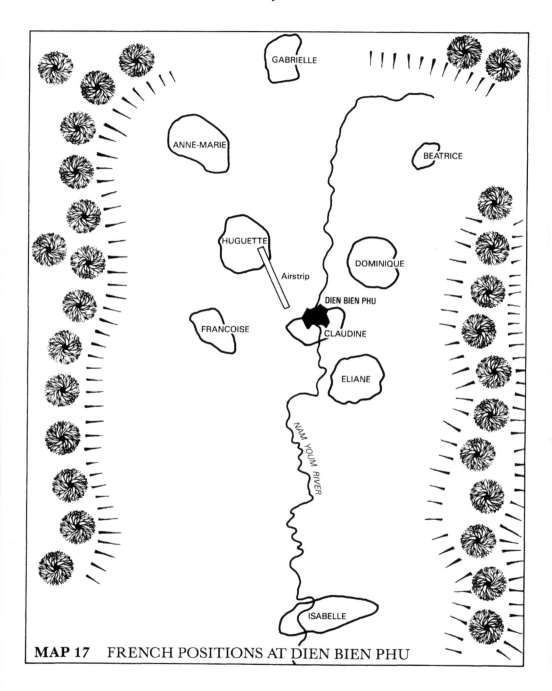

MAP 17   FRENCH POSITIONS AT DIEN BIEN PHU

M24 Chaffee light tanks of 1er Chasseurs à Cheval *respond to enemy fire during an ambush.* (ECPA)

Division from the west. The 351st Heavy Division also arrived with its artillery and engineer units, its journey to the front being one of the most remarkable in the annals of jungle warfare. Giap had interpreted the French strategy correctly from the outset and he was determined not only to crush his opponents' artillery with overwhelming firepower, but also to establish an anti-aircraft defence which would cripple attempts to reinforce and supply the besieged garrison by air. To this end he had assembled a 75,000-strong labour force of both sexes from the village militias and other sources to open trails through the forest. His heavy weapons were then dismantled and carried forward piece by piece, together with the vast quantities of ammunition required, using any sort of transport available, including the bicycle. Having arrived on the hills surrounding Dien Bien Phu, these were then dug in on the reverse slopes, where the jungle canopy made their presence almost undetectable. Altogether, the Viet Minh were able to deploy 144 field guns of 75mm or 105mm calibre, forty-eight 120mm mortars, thirty 75mm recoilless guns, twelve six-barrel *Kutyusha* rocket launchers, and over 180 anti-aircraft weapons of between 12.7mm and 37mm calibre, the last manned by a regular Chinese air-defence regiment.

Having achieved an overall superiority of 8:1, Giap began his assault on 13 March with a heavy bombardment. The main airstrip was hit and three aircraft destroyed. During the evening the 312th Division's human wave attacks swamped the defences of *Beatrice*, only 200 men of its Legion garrison managing to reach the main perimeter of the base. The following night it was the turn of *Gabrielle*, which managed to hold out until dawn, when a counter-attack with tank support enabled the 150 survivors to withdraw. During the night of 16/17 March the French sustained a further blow when the T'ai battalion defending *Anne-Marie* deserted; the effect of this was to expose the northern end of the main airstrip so that it was no longer usable by day.

The French High Command had begun reinforcing Dien Bien Phu with parachute battalions on 14 March, and continued to do so until the bitter end, when seven of the thirteen battalions present consisted of paratroopers. Once the airstrip became unusable it also para-dropped supplies, but because the perimeter had contracted, up to 30 per cent of these drifted into the Viet Minh lines. During the 55 days of the siege Giap's anti-aircraft units shot down or seriously damaged 62 aircraft. Air strikes with rockets and napalm failed to silence the enemy artillery, which made movement above ground impossible and inflicted three-quarters of the French casualties.

The first encounters had cost Giap some 2,500 men but he tightened his grip on the last outpost, *Isabelle*, and the inner perimeter, constructing trench lines until the battlefield resembled a sector of the Western Front during the First World War. On the night of 30 March the 312th and 316th Divisions launched a desperate series of attacks on *Dominique* and *Eliane*. *Isabelle* was also attacked by the 304th Division and on 2 April the 308th Division assaulted *Huguette*. Despite savage hand-to-hand fighting which lasted until 5 April, little progress was made. So horrific were the Viet Minh's casualties that for a while Giap faced a revolt against his brutal human wave tactics.

The battle, however, was not one he could afford to lose, for on 8 May an international conference convened by all the interested parties was due to discuss the question of Indochina in Geneva. He rested his troops and absorbed reinforcements while his artillery continued to hammer the garrison. Then, on 1 May, he returned to the attack. The French had only three days rations in hand but fought on until the evening of 7 May, when the last pocket of resistance was overwhelmed. Giap's victory had cost him a minimum of 8,000 dead and 15,000 wounded. French losses amounted to 2,293 killed and 11,000 captured, including 5,143 wounded, but only 3,000 of the prisoners survived their captivity; the remainder died during the march to the prison camps or as a result of being 're-educated' when they arrived.

Dien Bien Phu did not in itself end the war in Vietnam and the troops lost amounted to a mere 5 per cent of those which France had available, although they included the cream of Navarre's strategic reserve. On the other hand, the outcome of the battle placed the Viet Minh delegates at the Geneva conference in a very strong position. As early as November 1953 French Prime Minister Joseph Laniel had stated that 'If an honourable settlement were in sight, on either the local or the international level, France would be happy to accept a diplomatic solution to the conflict.' Ho Chi Minh did not trust the French and was prepared to exploit his recent victory until he could dictate terms. However, the Soviet Union and China, whose economies had suffered as a result of the Korean War and the need to support the Viet Minh in Indochina, urged him to be flexible; indeed, the Chinese, who were worried that the United States might intervene in Indochina, warned him that they would cut off his aid if he was not. From 23 June onwards, the only negotiations which counted took place between France's newly invested Prime Minister, Pierre Mendes-France, the Prime Minister of China, Chou En-lai, and the Soviet Foreign Minister, Vyacheslav Molotov. When the conference broke up at the end of July it was agreed that Laos, Cambodia and Vietnam would be granted full independence; that Vietnam would be temporarily partitioned along the 17th Parallel pending nationwide elections to be held in 1956; and that in the meantime the French and Viet Minh forces would withdraw respectively from north and south Vietnam.

CHAPTER 7

FIREPOWER VERSUS WILLPOWER

The Geneva agreement may have suited the needs of the Great Powers, but it satisfied few in Vietnam itself. In the north, Ho Chi Minh felt cheated of victory, resented the cavalier attitude of his patrons, and clearly did not regard the accord as being anything more than a truce. For the moment, however, he was faced with internal problems caused by the exodus of many thousands to the south, the majority of whom were members of the professions or skilled workers, and by the lack of access to Vietnam's major rice-growing area, the Mekong Delta. He could do little about the former, but in December 1955 he embarked on an extensive land reform programme, including the public trial of 'landlords', real or imaginary, and the imposition of such severe ideological solutions to the agricultural economy that risings took place in some areas. After thousands had died in the ruthless suppression of these, he commented sadly on his inability to 'wake the dead', but generously conceded that many of his measures had been inappropriate. Notwithstanding his doubtless sincere regrets, such events served to eliminate the last traces of opposition and enabled the Party to establish its iron discipline throughout the country.

South Vietnam was effectively ruled by Ngo Dinh Diem, who served first as Prime Minister to the Emperor Bao Dai, and then as President of the new republic after the latter was deposed following a referendum held on 23 October 1955. Diem, a devout Catholic, also set about eliminating his opponents, both political and religious, and announced that he had no intention of taking part in the nationwide elections required by the Geneva agreement, since such a process was impossible in the communist-controlled north. In accordance with the terms of the agreement, the French Army had remained in the south as a guarantor until 1956, and it entertained hopes that it would subsequently be required to train Diem's recently established Army of the Republic of Vietnam (ARVN). Diem, however, preferred American military aid and on 28 April 1956 an American Military Assistance Advisory Group was set up.

Having settled his domestic problems, Ho decided to renew the revolutionary war process in South Vietnam. In October 1957 communist guerrilla groups began operating in the Mekong Delta. Simultaneously, a supply route from North Vietnam was opened up, winding along jungle tracks in the border areas of neutral Laos and Cambodia, becoming known as the Ho Chi Minh Trail. Soon, the familiar campaign of terror and intimidation began in the villages, accompanied by the wholesale murder of local officials and increasing attacks on government troops and police. On 8 July 1959 two American military advisers were killed at Bien Hoa. By December 1960 the guerrillas had established themselves so firmly throughout the country that the National Liberation Front for South Vietnam was able to declare itself openly in Hanoi; it was referred to in the south as the Viet Cong (Communist Vietnamese), and by the Americans as the VC or, more commonly, Charlie.

During 1961 the problem showed every sign of worsening and in February 1962 the American military advisory group was expanded to become the US Military Assistance Command under General Paul D. Harkins, being joined by Australian advisers in August. To contain the rural violence, some of the methods employed by the British during the Malayan Emergency were tried, but with only limited success. A Strategic Hamlet programme was begun with a view to providing protection for the worst affected areas. Unlike the Chinese squatters

Right *APCs of the US 5th Division carry out a sweep near the Demilitarized Zone.* (USAMHI)

Below *The monsoon rains turn a harbour area of 11th Armored Cavalry Regiment into a sea of mud.* (USAMHI)

in Malaya, however, the Vietnamese villagers owned their land, had farmed it for generations, and regarded their removal to the defended compounds as some sort of punishment. Furthermore, their sense of pride made them resent having to live off government handouts until the first of the new crops could be harvested. Even worse, since many of the Viet Cong were themselves villagers, they were able to penetrate the Strategic Hamlets from the outset. Nor did offers of amnesty produce the hoped-for results as the guerrillas, far from being a racially separate group starving in the jungle, were living as part of the established community, farming by day and fighting by night.

Like the French before them, many American advisers believed that the elusive enemy could be defeated if only he could be brought to battle. That this was not neces-

sarily so was demonstrated by the action fought at Ap Bac in the Mekong Delta, 40 miles south-west of Saigon, on 2 January 1963. Here, a force consisting of three regular Viet Cong companies, reinforced with a number of regional guerrilla units, had been located. The Viet Cong commander was aware that the ARVN 7th Division was closing in for the kill and did not expect to survive the encounter, but he dug in his men along a mile-long stretch of the Cong Luong Canal between Ap Bac and the next village, Ap Tan Thoi, taking advantage of the waterside vegetation which provided concealment yet offered a good field of fire for his machine guns and mortars across open rice paddies.

The ARVN plan was for most of the 11th Infantry Regiment to be landed by helicopter at the northern end of the communist line, near Ap Tan Thoi, while two battalions

Tanks and APCs of I/1st Cavalry establish a stop-line during a search and destroy mission by the Americal Division, south-west of Duc Pho, October 1969. During the later stages of the war, for political reasons, the term search and destroy was replaced by reconnaissance in force. (USAMHI)

of Civil Guards advanced on Ap Bac from the south and a rifle squadron of the 2nd Armoured Cavalry Regiment, carried in M113 armoured personnel carriers (APCs), closed in from the west. The 11th Infantry were successfully air-landed at 07:30 but shortly afterwards the Civil Guards' attack broke down when their commander was wounded. The operation quickly lost its momentum as the Vietnamese bickered among themselves and disregarded the advice of their American advisers. At length it was decided that the Civil Guards would simply hold their ground while the helicopters, which had American pilots, were used to insert three reserve companies just west of Ap Bac. The Viet Cong held their fire until the last possible moment, hitting all but one on the 15 aircraft and shooting five of them down. The APC squadron, located approximately one mile to the west, declined to intervene for a further three hours. When it

finally advanced, it did so in piecemeal fashion, enabling the Viet Cong to concentrate their fire on each vehicle in turn, bringing the hesitant advance to a halt and killing no fewer than 14 of the APCs' .50-calibre machine gunners. Thus far, the task force could hardly be said to have done well, although in its clumsy way it had managed to box in the communists from the north, south and west. By mid-afternoon, however, the ARVN corps commander had unbelievably managed to snatch an absurd defeat from the jaws of victory. He had made a parachute battalion available but, despite the despairing pleas of his advisers, he insisted that the drop should be made *west* of the APC squadron, where it could do no possible good, instead of the east of the canal, where it would complete the encirclement of the Viet Cong. The drop itself was made at dusk and provided a final tragic touch of *opera bouffe* as ARVN paratroopers and in-

M113 APC of 17th Cavalry mounting a 106mm recoilless rifle, advances in support of II/3rd Infantry, 199th Light Infantry Brigade, during an operation in the Binh Chanh district, April 1968. (USAMHI)

fantrymen opened fire on each other in the gathering darkness. The fiasco at Ap Bac cost the ARVN 61 killed and 100 wounded; much encouraged, the Viet Cong escaped across the canal, leaving but three bodies behind. Some small good came of the affair in that gunshields were fitted to the M113s' .50-calibre mountings.

The ARVN had many better days than that at Ap Bac, and indeed some good days, but for the moment it remained, at the higher levels at least, an amateur army attempting to deal with a professional guerrilla force. By the end of the year it had the benefit of 15,000 American advisers but was losing the equivalent of one battalion a month in casualties and stood in real danger of being defeated. Nor was the unsettled political climate within South Vietnam of any assistance to the war effort. Diem, who had pursued a policy of harsh repression against the Buddhist population, was overthrown by a military coup on 1 November 1963 and murdered the following day. He was succeeded by a series of senior officers, none of whom enjoyed popular support. The root of the matter was that while most people were opposed to the communists, they were not prepared to back administrations which openly practised nepotism and political preferment while turning a blind eye to racketeering. It seemed to many Americans that everyone who was anybody in Saigon had an angle of some sort to exploit for his own ends. Nevertheless, the United States was committed to maintaining the South Vietnamese regime by virture of its own Domino Theory, which held that if South Vietnam fell to the communists, so too would Laos and Cambodia, followed in turn by Thailand, Burma and Malaysia.

During 1964 the slide towards a major conflict continued rapidly. On 20 June General William C. Westmoreland took over the Military Assistance Command and the number of American advisers rose to 23,000.

Early in August North Vietnamese torpedo boats mounted unsuccessful attacks on American destroyers in the Gulf of Tonkin. The US Seventh Fleet's carriers responded with air strikes against naval bases and other military targets in North Vietnam and Congress passed a resolution empowering President Lyndon Johnson to take whatever steps were considered necessary to repel attacks on American forces and prevent further aggression. In October two Americans were killed when the Viet Cong mortared Bien Hoa air base, and in November a terrorist bomb attack on an American billet in Saigon killed two more and injured 52. Further attacks in the New Year resulted in Operation Rolling Thunder, the sustained bombing of targets in North Vietnam, and on 8 March 1965 the arrival of two Marine Corps battalions at Da Nang airbase signalled the formal entry of the United States into the ground war. By the end of 1968 there were 536,000 American troops deployed in support of the ARVN, plus contingents from Australia, New Zealand, Thailand and South Korea.

The Second Vietnam War was unique in that it was fought without established lines and that the physical possession of territory counted for little. It has been described, in the simplest terms possible, as a contest between firepower and willpower. The Americans and their allies sought to destroy the Viet Cong and the North Vietnamese Army (NVA) units supporting them, using the ample technical resources at their disposal. On the other hand Giap, the communist commander, sought to destroy the Allied will to fight, knowing that he could never defeat the Americans in the field. As callous as ever of casualties, he commented that 'the life or death of a hundred, a thousand, tens of thousands of human beings, even our compatriots, is of no account. We shall fight for ten, fifteen, twenty, fifty years, regardless of cost, until we have won the final

Above *A returning patrol of II/9th Marines secures its weapons as it passes through the perimeter of its base at Bich Nam, near Da Nang, March 1966. The average age of American troops serving in Vietnam was 19, seven years younger than those who had fought against the Japanese in the Second World War. (USMC)*

Below *US 25th Division M106 4.2-inch mortar carrier on standby during operations near Phuoc-my, April 1966. Note the sandbag breastworks around the hatches, enabling the crew to use their weapons from inside the vehicle. (USAMHI)*

victory.' Nor was this mere rhetoric; Giap believed and meant what he said.

The French had fought in Vietnam using the weapons of the Second World War, but since then science and technology had advanced at a bewildering pace. Thus, the Second Vietnam War not only consolidated all the previous lessons of jungle warfare, but also involved the operational use of electronic and other devices as well as advanced weapon systems which led to the development of more sophisticated tactical techniques. Before describing the course of the fighting, therefore, it is necessary to examine some of these aspects in detail.

The use by the communists of neutral territory, darkness and the jungle to conceal their activities was countered in a number of ways. At the highest level photographic satellites, orbiting above Laos and Cambodia, were capable of sending down pictures of movement along the Ho Chi Minh Trail, while closer to home the remotely piloted AQM-34L, fitted with a 2,000-exposure camera and a television system which transmitted to an airborne receiving station, was used for distant on-the-spot reconnaissance, overflying the target area at 1,500 feet before climbing to 50,000 feet for the return flight. RF-4C Phantoms were also used for photo reconnaissance, equipped with forward-and side-looking cameras and infra-red sensors which would pick up any source of heat. Seismic sensors were used at both the strategic and tactical levels. At the strategic, they were air-dropped in the desired pattern, burying themselves in the jungle floor with only the antenna showing. They had a battery life of up to 45 days, during which they relayed seismic intelligence via a long-endurance aircraft to a distant Infiltration Surveillance Centre where, with computer assistance, target information was produced for air strikes. At the tactical level, they could be incorporated into the outer defences of a fortified base and connected by buried cable to the command post, giving early warning of the presence of intruders. Ground radar, image intensifiers and night sights enabled the troops to search the surrounding darkness, but were obviously more effective in open country than in the jungle.

One of the odder devices was the 'people sniffer', which was fitted first to a helicopter then to a specially designed light aircraft with a silenced engine, the YO-3A. The 'people sniffer' was sensitive to the gasses given off by decomposing faeces, even below the jungle canopy, and the principle was that the larger the sensor reading, the greater the number of Viet Cong hiding below. This could produce acute embarrassment, as it did when a spectacularly large reading led to the entire area being hammered by the corps and divisional artillery, air strikes and gunships. When the ground troops moved in, however, they found, not the dazed survivors of the Viet Cong division they had expected to meet, but the bodies of several elephants. As one disgusted participant recalled, the operation had produced the most expensive manure heap in history.

Another means by which the Americans sought to deny the Viet Cong the sanctuary of the jungle was defoliation. This involved the use of herbicides sprayed over selected areas by C-123 transport aircraft flying in formation. Several agents were tried, code-named White, Purple and Orange. Of these, Agent Orange was the most efficient, although the dioxin used has subsequently been linked with cancer and genetic defects. In some respects herbicides were a two-edged weapon in that they also killed crops and forced farming communities to move into the Strategic Hamlets, thereby depriving the guerrillas of food and support, but the resentment they generated also produced a steady flow of recruits for the Viet Cong, who exploited the propaganda advantage to the full. Alternative methods of jungle clearance included napalm, bombing and bull-dozing with giant Rome ploughs. Napalm produced poor results, as the jungle was too damp to support a self-sustaining forest fire; bombing and bulldozing were slow, expensive and actually encouraged fresh growth in the monsoon climate, so that the cleared areas were soon covered with dense secondary jungle.

Armoured vehicles were used on a greater scale than in any previous jungle war. As we have seen, the M113 APC, which possessed an amphibious capability and could be used to cross waterways and flooded paddy fields, was already in service with the ARVN before the first American units arrived. Experience revealed that if the infantry dismounted before the enemy's fire had been thoroughly suppressed they for-feited much of their advantage, and to correct this additional side-mounted automatic weapons and gunshields were fitted above the crew compartment. This arrangement was standardised in 1966 by the US 11th Armoured Cavalry Regiment and became known as the Armoured Cavalry Assault Vehicle (ACAV). Sometimes the armament was supplemented by an M79 automatic grenade launcher, and often sandbags were arranged around the edge of the troop compartment, serving as parapets for the infantry in action. Because of the Viet Cong's expertise in the use of mines, which could have a devastating effect in the crowded interior of the vehicle, the infantry preferred to ride outside unless in contact with the enemy.

The attitude of the Military Assistance

1st Squadron 11th Armoured Cavalry ACAVs open a destructive fire into suspected enemy positions near Long Binh, February 1969. (USAMHI)

Command to tanks was at first lukewarm, as it believed that they had no place in what were initially seen as counter-insurgency operations. The Marines, however, brought their organic tank units with them and the success of these, together with the recognition that the Allies were fighting a full-scale guerrilla campaign which also involved the North Vietnamese Army (NVA), led to the universal use of tanks in all but the most impossible terrain. The Americans used the M48 Patton series main battle tank and also evaluated the M551 Sheridan light tank, although the latter proved vulnerable to mine damage and suffered from technical teething problems; the ARVN used the M24 Chaffee and the M41 Walker Bulldog light tanks and, later, the M48 series; and the Australians used the Centurion. Two basic unit organisations were employed by the US Army, but in practice these varied widely. The first was the tank battalion, which con-

tained a headquarters company, a service company and three tank companies, each of three five-tank platoons. The second was the divisional armoured cavalry squadron, consisting of a headquarters troop, an air-mobile troop with helicopters, and three armoured cavalry troops each containing troop headquarters and three armoured cavalry platoons containing APC, ACAV and tank sections. In addition, there was one major formation, the 11th Armoured Cavalry Regiment, which consisted of a headquarters troop, an air-mobile troop and three armoured cavalry squadrons, each containing three armoured cavalry platoons with ACAVs, a tank company and a howitzer battery. As unit designations differ in their meanings between branches of the service and between armies it is necessary to explain that those applicable to Allied armour serving in Vietnam are open to misinterpretation. Thus, while Australian and ARVN

M48 tank supporting II/4th Marines engages targets south of the Demilitarized Zone, February 1968. (USMC)

Above *In 1969 the M551 Sheridan light tank began replacing the M48 medium in divisional armoured cavalry squadrons. The vehicle's unreliability and vulnerability to mine damage made it less than popular.* (USAMHI)

Below *Australian 1st Armoured Regiment Centurion batters its way along a jungle track. The characteristic side-plates were removed for this type of operation as foliage tended to compact between them and the tank's running gear. In such circumstances fuel consumption could rise to 12 gallons per mile.* (Australian War Memorial)

Above *Tanks of I/77th Armor form a stop-line during a US 5th Infantry Division search and destroy operation near Mai Loc. The I/77th was the last American tank battalion to serve in Vietnam. (USAMHI)*

Below *A column of Marine LVTP-5 prepares to move inland. The vehicle was amphibious and could be used as an APC, but was dangerously vulnerable to mines, RPG-7s and 75mm recoilless rifle fire. (USMC)*

units employed the terms regiment, squadron and troop, the US Armoured Cavalry's equivalents were squadron, troop and platoon, while the remainder of the US Army and Marine Corps armour preferred battalion, company and platoon. It can, therefore, be seen that the US 11th Armoured Cavalry Regiment was over three times the strength of an Australian or ARVN armoured regiment.

The principal roles in which armoured troops were employed included search-and-destroy missions, rapid reaction, line of communications security and perimeter defence. The search-and-destroy concept was very similar to that of the British cordon-and-search in that the suspect area was surrounded and likely avenues of escape sealed by cut-off parties or artillery fire; the area was then subjected to one or more sweeps which drove the enemy towards the stop-line and revealed his positions. Ever since the introduction of hand-held anti-tank weapons during the Second World War it had been customary for the infantry to lead in close country, but because of the Viet Cong's use of anti-personnel mines, booby traps and ambush techniques this was reversed. Both the M48 and the Centurion were powerful enough to break new trails through the forest, detonating mines and eliminating booby traps, and were capable of absorbing considerable punishment from the enemy's RPG-7 anti-tank rocket launcher. During their advance the tanks raked the jungle with their automatic weapons, as did the mechanized infantry following close behind in their APCs, thereby discouraging ambush parties. Once the enemy bunkers had been reached they could

M48 of Company B 1st Marine Tank Battalion laden with C rations and extra ammunition for III/7th Marines. Beneath the boxes can be seen a girdle of track links which protect the turret against RPG-7s. (USMC)

be smashed up with armour-piercing and high explosive ammunition, then crushed beneath the tracks. Against troops canister was used, the effect being to turn the tanks' main armament into a huge shotgun; later, the canister round was replaced by Beehive ammunition, which contained thousands of small metal darts and could be fused to explode from point-blank out to 4,000 yards. Both types of round could also be used to strip vegetation from suspected positions, to clear barbed wire and to explode mines from the path ahead. If the enemy made a stand, tanks and APCs would engage him closely while additional artillery and air strikes were laid on and fresh troops lifted into the area by helicopter, a technique known as the Pile On. The final phase of the operation would consist of a dismounted sweep by the infantry, who would search the position for prisoners and intelligence material and destroy captured weapons and equipment.

Rapid reaction involved a force of tanks and/or APCs remaining on standby within a base, ready to go to the assistance of a road convoy or patrol which had run into trouble. Line of communication security consisted of route opening and convoy escort, and accounted for approximately one-third of all armoured operations. The aim of the guerrillas was to deny free use of the roads between military garrisons and their supply

1st Squadron 11th Armoured Cavalry ACAVs and tanks probe into an enemy held village after heavy artillery and air strikes. (USAMHI)

M113 of the Australian 3rd Cavalry Regiment, fitted with Cadillac-Gage T-50 turret. (Simon Dunston)

bases and to levy contributions from the civilian traffic, just as they had in the days of the French, and as the degree to which roads were free from interference directly reflected the extent of their control over the surrounding countryside, these activities accounted for a considerable part of their total effort. Ambush sites were approximately half-a-mile in length, the ambush being triggered by command mines which halted the leading and last vehicles. A heavy fire was then opened with automatic weapons, mortars, RPG-7s and recoilless rifles, the emphasis being on inflicting heavy casualties; those who attempted to take cover on the roadsides or in ditches found them littered with anti-personnel mines, *panji* stakes and booby traps. Sometimes the real target of the ambush was not the column itself but the rapid reaction force, which would run into a second and more deadly ambush on its way to the site. Having completed its task, the ambush team would then disperse rapidly.

The counter-ambush technique in-volved tanks, ACAVs and APCs swinging alternatively right and left to the roadside in a herring-bone pattern and opening fire with every weapon available for what was described as the 'mad minute'. The sheer volume produced stripped the undergrowth in which the guerrillas were concealed and so enabled more accurate aim to be taken. If possible, the leading AFVs would then drive through the ambush and 'clover-leaf' round onto the enemy's rear, a response which, it will be recalled, had been recommended by General Sir Frederick Roberts in Burma during the 1880s. In the event of the ambush site being a long one, troops would pass through each other, alternately adopting the herring-bone pattern until clear of the danger area. While the fighting was in progress unarmoured vehicles would continue to drive forward, pushing disabled hulks aside. Most ambushes were set at night, and to discourage the practice groups of AFVs would make irregular but continuous patrols at speed, raking the road-

163

sides and likely ambush sites with their fire; patrols of this nature were generally referred to as 'thunder runs'.

When armoured units were not engaged in active operations, particularly during the wet season, their vehicles were incorporated into the perimeter defence system of whichever base they were operating from. The vehicles and their crews were dug in to protect them from the mortar or rocket bombardments which normally preceded a communist attack and chain link fencing was used as a defence against RPG-7s. Arcs of fire were interlocked, the combined potential of the AFVs being used to discourage attacks with a 'mad minute' of rapid fire during the last light stand-to. After dark,

Left *Centurion bridgelayer of B Squadron 1st Armoured Regiment drops its bridge across a jungle water obstacle.* (Simon Dunston)

Below *The combination of mud and AFVs made life difficult for the inhabitants of any base, but the latter were welcome additions to defensive firepower. The stand-by turret crew of this Centurion have fitted a sunshield to protect themselves from the worst of the day's heat.* (Simon Dunston)

the tanks' Xenon infra-red light projectors provided a welcome source of illumination when required.

The communists made occasional use of the Soviet PT-76 light amphibious tank or its Chinese equivalent the Type 63 during attacks on isolated base camps, but these were seldom successful. During the final phase of the war, however, they used T-55 and Type 59 medium tanks in a conventional offensive.

Artillery also played a greater role than in previous jungle wars. The principal Allied weapons were the veteran M101A1 towed 105mm howitzer, progressively replaced from March 1966 onwards by the M102 towed 105mm howitzer, which had a wider traverse and was lighter, enabling more ammunition to be carried on airmobile operations; the M108 self-propelled 105mm howitzer, which was technically obsolete although still capable of making a contribution; the M114A1 towed 155mm howitzer, also obsolete but retained because of its air-

mobile potential; the M109 self-propelled 155mm howitzer; the M107 self-propelled 175mm gun; and the M110 self-propelled 8-inch howitzer.

The cornerstone of Allied artillery tactics in Vietnam was the fire base, the primary purpose of which was to provide artillery support for friendly troops operating in the area. In addition to the artillery personnel, the fire base contained an infantry element which was responsible for its defence. The base was commanded by the senior artillery or infantry officer present, but its site was jointly agreed by the artillery and infantry. The former were concerned with their ability to give all-round fire support, the ability of neighbouring fire support bases to provide indirect support if the base was attacked, the capacity of the ground to support the guns, and whether air supply was possible; the latter were concerned with the defensibility of the site and its central location in relation to the area in which they would be operating. In a typical fire base the

M107 175mm self-propelled gun firing in support of 7th Marine Regiment from a fire base at Hill 55, March 1969. (USMC)

Left *Marines at a fire support base south-east of Da Nang receive additional supplies from a CH-53 Sea Stallion helicopter. The weapon is the tried and tested M101 105mm howitzer, the first versions of which entered service prior to the Second World War.* (USMC)

Below right *Lowering an M102 105mm howitzer into a prepared gun position at a fire support base.* (USAMHI)

Below *Chinook helicopter arrives above a recently cleared landing zone with an M102 105mm howitzer and ready-use ammunition.* (USAMHI)

six-gun batteries, which included all 105mm- and 155mm-equipped units, were emplaced in a five-pointed star with five guns at the points of the star and the sixth in the centre; in the event of a night attack the sixth gun would fire illuminating rounds while the remainder engaged over open sights. A diamond pattern was adopted by composite 175mm and 8-inch batteries, with the 175mm guns located furthest from the command post to reduce the blast effect. Perimeter defences were established by the infantry, who dug trenches and bunkers, strung wire, set up Claymore mines and trip flares, and emplaced their 81mm and 4.2-inch mortars, which fired both high explosive and illuminating ammunition. Ideally, the base defences also included several dual- or quadruple-barrel automatic anti-aircraft weapons, which were prized for their high output against ground targets, although such targets were not always available. Continuous aggressive patrolling around the

base inhibited enemy activity in its immediate vicinity.

Once the guns had been emplaced the fire base could respond instantly to calls for artillery support, including those from neighbouring fire bases. Counter-mortar fire was unobserved and was either preplanned, using knowledge of the enemy's weapons and knowledge of the terrain to predict probable base-plate positions, or assisted by mortar-locating radar. Opensights firing was employed in direct defence of the base, using Beehive rounds or, when the enemy learned to avoid the worst effects of this by lying down or crawling, high explosive fused to air-burst 30 feet above the ground at ranges from 200 to 1,000 yards. The latter technique was referred to as Killer Junior when employed by 105mm and 155mm weapons, and as Killer Senior when fired by the 8-inch howitzer.

Because of the fire base system there were very few occasions when Allied troops

operated beyond the range of supporting artillery, and this in itself made the bases prime targets for Viet Cong and NVA attacks. One such attack was launched by two companies of the NVA's 22nd Regiment against Landing Zone Bird, 50 miles north of Que Nhon in Binh Dinh province, during the night of 26 December 1966. Bird contained a fire base held by Battery B II/19th Artillery with six 105mm howitzers, Battery C VI/16th Artillery with six 155mm howitzers, and the half-strength Company C II/12th Cavalry. Under cover of darkness the communists crawled to within feet of the perimeter. Shortly after midnight a rain of mortar shells hit the position and the northern defences were swamped by a rush of NVA infantry. The defenders gave ground slowly but were pressed steadily back. At the critical moment two rounds of 105mm Beehive were fired into the thick of the attackers

and survivors quickly melted into the darkness. The Americans lost 30 killed but 266 NVA bodies were counted in and around the base.

An excellent example of integrated defence was provided during the attack on Fire Support Base Burt, located 6 miles from the Cambodian border in Tay Ninh province, during the night of 1/2 January 1968. The base was held by Battery A III/13th Artillery with 155mm self-propelled howitzers and Batteries A and C II/77th Artillery with 105mm howitzers, perimeter defence being provided by II and III Battalions 22nd Infantry (Mechanized). Mortaring had begun during the afternoon and shortly before midnight a diversionary attack was mounted against the western perimeter. The main attack was then launched against the southern perimeter and both 105mm batteries commenced fir-

Permanent fire support bases were surrounded by chain link fencing which served as a defence against the enemy's rocket-propelled grenades. (USAMHI)

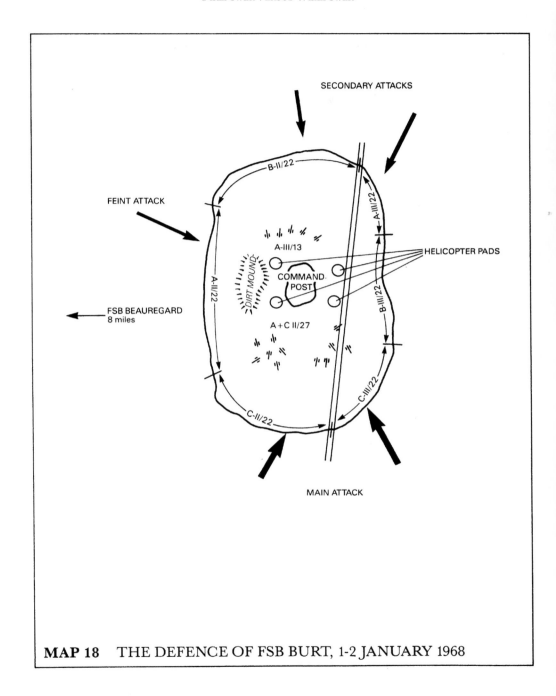

MAP 18   THE DEFENCE OF FSB BURT, 1-2 JANUARY 1968

Above *M114 155mm howitzer of Battery G 77th Artillery firing in support of 1st Cavalry Division in the Bong Son district, February 1966. The breastwork surrounding the gun position consists of expended ammunition cases filled with earth and topped by sandbags.* (USAMHI)

Below *Battery of M101 105mm howitzers firing in support of 1st Cavalry Division operations, August 1969.* (USAMHI)

ing over open sights with Beehive and Killer Junior, assisted by flares dropped from aircraft and armed helicopters. On the northern sector the 155mm battery was also engaging over open sights. Supplementary fire was requested from Fire Support Base Beauregard, some eight miles from the west, and this was used to protect the western perimeter. Later, Burt's 155mm battery and Beauregard's guns switched to indirect fire against the road running south from Burt, where it was believed, correctly, that the main body of the enemy force was concentrated. Tactical air support became available at 03:00 and continued to pound the area to the south; by 06:00 the enemy had gone. The Viet Cong had employed four battalions in the attack, which cost them 400 dead. American casualties in the action, officially known as the Battle of Soui Cut, amounted to 23 killed and 153 wounded.

A classic example of the aggressive use of the fire base was provided by the Battle of Long Tan. On 17 August 1966 the First Australian Task Force base at Nui Dat in Phuoc Tuy province was subjected to a mortar attack. The following day D Company 6th Royal Australian Regiment set off to patrol the area where the enemy mortars had been located. At 15:40 the company was moving through the rubber plantation at Long Tan, some two-and-a-half miles east of the base, when the leading platoon encountered several guerrillas. It gave chase but was itself halted by heavy fire and it was immediately clear that D Company had run into a much superior enemy force. The communists counter-attacked, forcing the company into defensive perimeter and, despite their fierce resistance, it is probable that the Australians would have been overrun had it not been for the guns of 1st Field Regiment Royal Australian Artillery, their fire from Nui Dat being expertly controlled by D Company's Forward Observation Officer, Lieutenant Maury Stanley, a New Zealander. The Aus-

A view of Fire Support Base Gettysburg in the desolate Plain of Reeds near the Cambodian border. (USAMHI)

tralian artillery consisted of 103, 105 and 161 (New Zealand) Field Batteries, all equipped with Italian-designed 105mm light howitzers, and Battery A, II/35th US Artillery, with M109 155mm self-propelled howitzers. Official ammunition expenditure during the action is quoted as 2,639 rounds of 105mm and 155 rounds of 155mm, but the guns were in action continuously from 16:19 and some participants believe that as many as 1,000 more rounds were fired. The overheated gun barrels began to steam in the constant rain, which also prevented the dispersion of expended cordite fumes and made breathing difficult. Volunteers from the nearest units arrived to assist the exhausted gunners by moving the mounds of empty cases and carrying ammunition when the re-supply helicopters arrived at 18:00. Meanwhile, a relief force consisting of 3 Troop 1st Armoured Personnel Carrier Squadron with

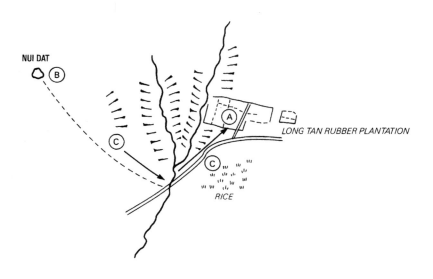

NUI DAT

LONG TAN RUBBER PLANTATION

RICE

A   D Company 6th Royal Australian Regiment isolated
     in rubber plantation
B   Artillery support from 1 Australian Task Force base
     at Nui Dat
C   Relief force consisting of A Company 6th Royal
     Australian Regiment and 3 Troop 1st APC Squadron

MAP 19   LONG TAN, 18 AUGUST 1966

The recently introduced M102 105mm howitzer was a technically superior weapon to the older M101, but its users found the lower breech extremely tiring during prolonged fire support engagements. (USAMHI)

A Company 6th Royal Australian Regiment aboard its M113s had left Nui Dat at 17:40. It reached Long Tan shortly before last light, passing through the defensive barrage, and swept through the plantation to rout the Viet Cong, some of whom were run over while others were cut down by the carriers' weapons. As the survivors attempted to withdraw, the artillery's fire was shifted onto their probable avenues of escape. Officially, the four Viet Cong battalions engaged in the action suffered the loss of 245 killed, but many more dead bodies were found in the area after the formal count had ended. Given the intense nature of the fighting in the plantation, Australian casualties were incredibly light.

During the early stages of the war the communist forces relied on recoilless rifles and mortars for their attacks on Allied positions, but the growing sophistication of mortar-locating and counter-battery tech-

niques led to the mortar being gradually supplanted by rockets of various calibres, which were easily transported, simple to launch, had a longer range and carried a heavier payload, although their lack of accuracy made them suitable only for use against area targets. By the end of 1969 the rocket had become so popular that rocket regiments had been formed. The major disadvantage of the rocket was that it left a trail for all to see and, for the sake of survival, it was necessary for rocket troops to change position after firing a very few rounds. Until the invasion of the south during the final stages of the war, the communists made little use of field artillery weapons save in the so-called Demilitarized Zone dividing North from South Vietnam, where artillery regiments were equipped with Soviet guns varying in calibre between 85mm and 152mm.

Another, and most important, factor which made the Second Vietnam War different from earlier jungle conflicts was the employment of helicopters *en masse*, so adding a third dimension to the battlefield. A few helicopters had flown operationally during the Second World War, but rotocraft first saw extensive military use during the Korean War, the First Vietnam War and the Malayan Emergency. The principal defect of the helicopters of that period was their poor power-to-weight ratio, which seriously restricted the weight of the load to be carried as well as the internal space available. This problem was solved by the introduction of the turboshaft engine in 1955 and thereafter it became possible to design helicopters for every conceivable military requirement, including reconnaissance craft, troop transports, ground attack gunships, cargo carriers and flying cranes. From this moment, the influence of the helicopter on the land battle was never in doubt, since it offered a precision and flexibility superior to that of airborne operations yet possessed a degree of firepower equal to that of many AFVs. Specifically, airmobility enabled a ground formation and its supporting weapons to be

lifted quickly into a target area, irrespective of intervening terrain factors, maintained there for the duration of the operation, then extracted in the same manner.

The first major airmobile operation of the Second Vietnam War, codenamed Chopper, took place on 23 December 1961 when 32 twin-rotor Vertol H-21 'Flying Bananas' with American pilots lifted a force of ARVN paratroops onto a Viet Cong headquarters complex 10 miles west of Saigon, routing the startled defenders and capturing a clandestine radio station. The success of further airmobile operations quickly convinced the Viet Cong and the NVA that the rules of the game had changed somewhat, as the following extract from a pamphlet, captured in November 1962, reveals:

Although we have succeeded in inflicting some loss on the enemy in his heliborne operations the enemy has in some places caused us fairly heavy losses. We must therefore find means of coping with the enemy's helicopter tactics. Widespread effort must be directed to combating heliborne landings and shooting at helicopters. Following are the advantages

which the enemy enjoys due to his employment of heliborne strike tactics:

1. Careful planning and preparations are possible together with complete mobility in an attack, support or relieving role.
2. Secrecy can be preserved and surprise strikes can be accomplished.
3. Landings can be affected deep into our rear areas with the capability to attack and withdraw rapidly.
4. An appropriate means of destroying our forces while they are still weak.

The most important elements of the helicopter fleet which the United States deployed in Vietnam were the Bell UH-1 Iroquois series, better known as the Huey, which served in a variety of roles but was most frequently employed as an armed troop transport with an eight-man infantry section aboard; the Bell AH-1 Huey Cobra gunship, which served as an escort and fire support craft and was armed with a variety of weapon systems, including one or two six-barrel miniguns capable of ripping a target to pieces with their high rate of fire, one or two 40mm grenade launchers, or pods containing seventy-six 2.75-inch rockets; the twin-rotor Boeing-Vertol CH-47 Chinook transport, capable of carrying 33 troops or their equivalent weight in cargo, plus a field piece or stores slung from a ventral hook; the Sikorsky CH-64 Tarhe flying crane, which could lift heavy artillery weapons and other awkward loads as well as recover crashed aircraft; the Bell OH-58A Kiowa and the Hughes OH-6 Cayuse light observation and reconnaissance helicopters.

The US Army had given much thought to the question of airmobility at the higher levels and this resulted in the formation of the 1st Cavalry Division (Airmobile), which began arriving in August 1965. The division had a strength of 16,000 men and possessed no fewer than 428 rotocraft. Its internal organisation was flexible and contained three brigade headquarters, eight infantry battalions, an aviation group, an artillery group which included three 105mm howitzer battalions and helicopter gunship units, signals and engineer battalions and divisional services. It was therefore possible for the divisional commander to allocate the appropriate number of infantry battalions

Left *Airmobile ARVN troops fan out after being lifted into position. Few landing zones were as convenient as this.* (USAMHI)

Right *Troops board a flight of Huey helicopters for an airmobile operation.* (USAMHI)

and supporting arms to individual brigade headquarters in accordance with the requirements of the mission in hand, after which they reverted to divisional control. The concept proved so successful that in June 1968 it was decided that the 101st Airborne Division, which had reached Vietnam the previous December, would be reorganized as an airmobile formation. A third formation, the 173rd Airborne Brigade, which ultimately contained four parachute battalions and supporting artillery, had become involved in airmobile operations as soon as it arrived in Vietnam in May 1965 and was in effect permanently airmobile, although it retained and occasionally used its airborne capability. In addition to these specialist formations, the majority of American and Allied infantry formations became progressively familiar with airmobile techniques as the war continued.

Once on the ground, the airmobile soldier fought as a conventional infantryman. The art lay in the manner of his insertion and extraction. During these periods he was at his most vulnerable and required support from artillery, gunships and tactical ground-attack aircraft, the activities of which had to be carefully co-ordinated so that the troop-carrying helicopters entered and left the landing zone (LZ) along lanes which lay clear of their trajectories. An LZ might be 'hot,' i.e. in enemy hands, in which case intense effort would be made to suppress the opposition by fire before the first troops landed, or it might be 'cold,' i.e. unoccupied, enabling the landing to be made immediately; often, no really reliable means existed of telling whether an LZ was 'hot' or 'cold' until the helicopters arrived overhead. The jungles of Vietnam were no more suitable for airmobile insertions than those of Malaya and if no clearing suitable for an LZ existed it was necessary to make one. This could be done by bombing or focussing artillery fire on the chosen site until sufficient timber had been felled. The first wave of infantry would then land and secure the perimeter while the second wave, equipped with LZ Kits containing explosive charges, axes and chain saws, set about enlarging the area. Extractions could be complicated by the enemy's presence; the following account by Brigadier-General Ellis W. Williamson,

Left *Flight of Hueys skimming the jungle canopy towards their objective.* (USAMHI)

Right *1st Cavalry Division Hueys commence their descent towards an LZ; the fact that few of the door guns are manned suggests this is already in friendly hands.* (USAMHI)

commanding 173rd Airborne Brigade, describes the end of a series of successful operations mounted north of the Song Dong Nai River in July 1965.

In all candour I must admit that I did not expect to find as many enemy in the area as we did. At the extraction time, we took 3,000 troops out of three different landing zones in three hours and ten minutes. We wouldn't have moved troops that fast and couldn't have afforded to bring our troops that close together unless we had had a lot going on simultaneously. As I looked at it from above, it was a sight to see. We were withdrawing from the centre LZ while some friendly troops were still in the western LZ. We had a helicopter strike going in a circle around the centre LZ. The machine gun and rocket-firing helicopters kept making their circle smaller and smaller as we withdrew our landing zone security. Just to the west side we had another helicopter strike running north to south. We also had something else that was just a little hairy, but it worked without question—the artillery was firing high angle fire to screen the northern side of the landing zone. The personnel lift helicopters were coming from the east, going under the artillery fire, setting down in the LZ to pick up troops, and leaving by way of the south-west. In addition to that,

we had an air strike going on to the north-east.

The airmobile version of the Rapid Response Force was known as the Eagle Flight. This might consist of one command helicopter, seven troop-carrying Hueys, five Cobras for escort and fire-support tasks and one casualty evacuation Huey. Eagle Flights remained on standby and were employed to bring enemy groups to battle once they had been located, or pin them down if they were attempting to break contact and withdraw. Such operations required the minimum of planning and were simply scaled-down versions of the larger airmobile operations. By November 1964 every helicopter company in South Vietnam maintained an Eagle Flight on permanent alert.

The first major battle involving airmobile troops took place in the Ia Drang Valley, Central Highlands, during October and November 1965. Here, American Special Forces teams, better known as the Green Berets, were performing the same function as had the British SAS in Malaya and Borneo. They lived among the *Montagnard* people, who were opposed to the communists,

1st Cavalry Division troopers saddle up prior to lift-off from their base camp. (USAMHI)

and organised Civilian Irregular Defence Groups (CIDGs) which operated from defended camps and preyed on the NVA's lines of communication. By the beginning of October they had become a thorn in the side of General Chu Huy Man, commanding the Western Highlands Field Front, and he decided to eliminate the more important Special Forces Camps in his area using three NVA regiments, the 32nd, 33rd and 66th, reinforced with Viet Cong battalions.

On 19 October an attack on the Special Forces Camp at Plei Mei was repulsed after a stiff fight. The 1st Cavalry Division, commanded by Major-General Harry Kinnard, was ordered forward from its base at An Khe and its leading elements assisted the ARVN units relieving Plei Mei. It was apparent that the enemy had retired westwards and 1st Brigade was involved in several clashes which confirmed this. On 9 November 1st Brigade was relieved by 3rd Brigade, com-

manded by Colonel Thomas W. Brown. Shortly after, Kinnard directed Brown to search the heavily wooded area covering the lower slopes of the Chu Pong massif, south of the Ia Drang.

The I/7th Cavalry, commanded by Lieutenant-Colonel Harold G. Moore, was detailed for the mission on 13 November. The battalion was to secure its LZ with an airmobile assault the following morning and carry out search-and-destroy patrols for the next two days. During the briefing Brown informed Moore that he would have the support of Batteries A and C I/21st Artillery, firing from Landing Zone Falcon, some eight miles west of Plei Mei; one battery was already present and the second was being lifted into position.

Moore had the choice of three LZs, codenamed Tango, Yankee and X-Ray. From the information obtained by his reconnaissance flights early on 14 November he

178

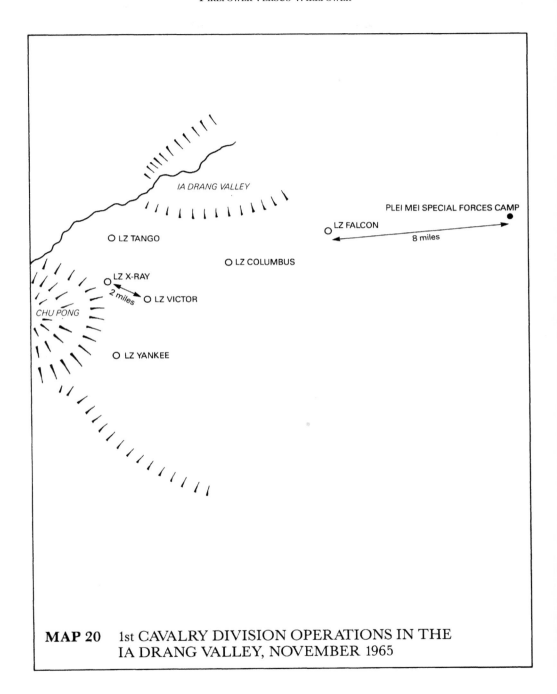

MAP 20   1st CAVALRY DIVISION OPERATIONS IN THE
IA DRANG VALLEY, NOVEMBER 1965

selected X-Ray as being the most suitable. At 10:17 the artillery and gunships opened fire on all three LZs to keep the enemy guessing. At 10:30 Company B completed its short flight from Plei Mei and touched down. Shaking out, its platoons began to advance up the slopes of Chu Pong. They soon became involved in a fierce fire fight with a large enemy force and one platoon was isolated. Company A moved up in support but was soon pinned down as well. By the time Companies C and D arrived the LZ was under automatic and mortar fire which was for the moment so intense that Moore used his radio to stop the last eight helicopters from landing. It was clearly apparent to him that the enemy was present in overwhelming strength and he took the only course possible by pulling back Companies A and B and establishing a defensive perimeter around X-Ray. Informed of the situation, Brown despatched the only reinforcements immediately available, Company B II/7th Cavalry, which was landed in the clearing at 18:00, by which time the enemy's fire had slackened somewhat. With the reinforcements came a pathfinder team

which set up landing lights under fire. By 19:15 additional ammunition, food, water and medical supplies had been flown in and those of the wounded that could be reached were lifted out.

A pall of smoke hung above X-Ray in the failing light. Moore's men had been inserted into a hornets' nest, for on the slopes of Chu Pong the NVA and Viet Cong were re-assembling for a fresh assault on Plei Mei. The arrival of I/7th Cavalry had given General Man a golden opportunity to inflict heavy casualties on the Americans and he took it with both hands. The 33rd and 66th Regiments and a Viet Cong battalion all converged on the perimeter, which was heavily probed throughout the night. At Landing Zone Falcon the sweating gunners fired over 4,000 rounds of 105mm high explosive in support of X-Ray, and a tactical air strike was delivered by the light of flares, but still the communists came. With the coming of dawn their attacks reached a new pitch of intensity and the artillery fire was called in to within 50 yards of the defenders. Even so, a breakthrough was averted only by Moore committing his last reserves. In places there

was savage hand-to-hand fighting. One platoon commander was found dead in his foxhole, surrounded by the bodies of five of his enemies; nearby, a trooper had been killed, his fingers locked around the throat of a North Vietnamese soldier. At 07:55 Moore ordered all platoons to throw coloured smoke grenades to mark the perimeter for incoming air strikes. These were delivered so close by helicopter gunships and fixed-wing aircraft that two napalm tanks fireballed inside the defences, fortunately without causing casualties. The slopes of Chu Pong itself erupted under the impact of hundreds of bombs dropped from B-52s—this being the first occasion on which these formidable heavy bombers were used in a tactical support role. The attacks faltered and faded away. Moore requested additional reinforcements and Brown, who was concentrating all his brigade's efforts on the relief of X-Ray, responded. At 09:00 Company A II/7th Cavalry were lifted in to take some of the strain off the weary garrison.

Late the previous afternoon Brown had lifted II/5th Cavalry, commanded by Lieutenant-Colonel Robert B. Tully, to Landing Zone Victor, two miles south-east of X-Ray, with orders to reinforce Moore's battalion the following morning. Tully's men had to fight most of their way forward but reached X-Ray at noon. During the afternoon the cavalrymen mounted a counter-attack and by 15:10 had broken through to the isolated platoon of Company B I/7th, which also stood off attacks throughout the night. The platoon commander and his sergeant had been killed and it had been commanded throughout its ordeal by a squad leader, Sergeant Clyde E. Savage, who had made expert use of supporting artillery fire. When the relief force arrived all but seven of its members were dead or wounded.

Colonel Brown also established a new fire base at Landing Zone Columbus, halfway between X-Ray and Falcon, using Battery B I/21st Artillery and Battery C II/17th Artillery. These redeployments meant that X-Ray was now defended by twice the number of infantry and had twice the artillery support. When attacks on the perimeter were renewed that night, therefore, they met even more determined resistance. At first light on 16 November the enemy broke con-

Left *A Chinook extracts cavalrymen and their heavy weapons from a hilltop LZ which has been cleared of trees and scrub.* (USAMHI)

Right *Trooper with a head wound receives assistance from medics. Thanks to an efficient casualty evacuation drill, most casualties received early treatment, a fact which in itself saved many lives.* (USAMHI)

tact and drifted away, enabling companies to patrol forward for a distance of 500 yards. A total of 634 bodies were counted and many more dead and wounded had obviously been dragged away. American casualties amounted to 79 killed and 121 wounded. Fighting in the Ia Drang Valley would continue for a further ten days but General Man's command had been decimated and he had learned the hard lesson that the unsupported courage and determination of his men could not hope to prevail against the firepower and mobility of airmobile troops.

Tactical air support was provided by a wide variety of jet- and piston-engined fixed-wing aircraft, armed with cannon,

Left *Good communications at all levels were vital in winning tactical victories.* (USAMHI)

Below *Booby traps on trails such as this meant that survival depended on vigilance. Most at risk was the man who marched point.* (USAMHI)

rockets, bombs and napalm. A request from troops for air support would be relayed by corps headquarters to the nearest air base or aircraft carrier. The aircraft would be scrambled and, once they had entered the target area, come under the control of the Forward Air Controller, who was either positioned with the ground troops or in radio contact with them from a Cessna O-1 Bird Dog, O-2 or North American OV-10 Bronco light observation plane. The target would be indicated either by air-launched rocket, artillery smoke round or verbal directions and the aircraft would make their attack runs. Most air strikes were made within 40 minutes of their being requested. As we have seen, the Boeing B-52 Stratofortress heavy bomber, which carried a 60,000-lb bom-

bload, was sometimes used tactically against large enemy concentrations. At night 200,000-candlepower parachute flares were used to illuminate a target area. In these circumstances, however, fast-flying jet aircraft, summoned to the assistance of an outpost under attack, had little time to identify their targets, and to remedy this an entirely new weapon system was devised by Captain Ronald Terry of the US Air Force Aeronautical Systems Division. Three Gatling-type six-barrelled 7.62mm miniguns, each with an output of up to 6,000 rounds per minute, including 25 per cent tracer ammunition, were fitted to fire out of the port side of the fuselage of a Douglas C-47 transport. Over the target area the aircraft banked to the left and circled slowly, bringing the guns to bear

A Douglas A-1 Skyraider of the South Vietnamese Air Force drops napalm canisters onto a target indicated by a forward air controller. The propellor-driven Skyraider had a maximum speed of 318 mph and was more vulnerable than jet attack aircraft; on the other hand, it had a longer flight endurance and carried up to 8,000-lb of external ordnance, plus four 20mm cannon. (US Air Force)

Above *The North American F-100 Super Sabre could carry a wide variety of ordnance but is here despatching a Snake-eye retarded-fall bomb during a low-altitude pass over the target area.* (USAF)

Below *Napalm tanks tumble from an F-100 Super Sabre during an attack on a Viet Cong concentration 21 miles east of Ban Me Thuot. The target was engaged and destroyed by a flight of four aircraft.* (USAF)

US Navy McDonnell Douglas A-4C Skyhawk from the attack carrier USS Bon Homme Richard *drops three 250-pound bombs on a suspected Viet Cong stronghold, November 1965.* (USAF)

in the light shed by a falling flare. Normally, three-second bursts were fired, sufficient to put one round into every square foot of an area the size of a football pitch. The aircraft was standardized as the AC-47 in 1965, and although it was often called Spooky its more common name was Puff the Magic Dragon. Later models were fitted with sensors and image intensifiers to penetrate the darkness. Subsequently, Fairchild C-199 and Lockheed C-130 transports were also converted to the gunship role, armed with a variety of weapons. When not directly involved in fire support missions, Puff often flew in company with psy-war aircraft which used their speakers to encourage the Viet Cong to desert; if the enemy responded with anti-aircraft fire, Puff would belch tracer at its source to emphasize the argument.

In addition to supplying direct tactical support for the ground troops, much of the Allied air effort in South Vietnam was directed at interdiction of the enemy's supply routes. This involved attacks on supply dumps, troop concentrations and lines of communication far from the battlefield. The use of bombs against bridges, gorges and mountain passes along the Ho Chi Minh Trail produced only limited success as the North Vietnamese maintained a large labour force which could fill in craters and restore the road in a matter of hours. Better results against truck convoys were obtained using sensor-equipped gunships at night. That sufficient supplies and reinforcements

Air-to-air view of a Cessna A-37 Dragonfly light strike aircraft of the 604th Special Operations Squadron releasing Snake-eye bombs over the target area, 2 September 1970. The Dragonfly became the principal strike aircraft of the South Vietnamese Air Force. (USAF)

Above *A Martin B-57 light tactical bomber releases part of its bomb load against concealed enemy positions in the Central Highlands. The B-57 was based on the English Electric Canberra and was built under licence in the United States. (USAF)*

Below *Bombs cascade from the belly of a Strategic Air Command Boeing B-52 Stratofortress. The B-52 had a bombload of approximately 60,000-lb. Tactical strikes by these aircraft were codenamed Arc Light and their effect was similar to that of a small earthquake. (USAF)*

Fast patrol craft of the type used to interdict the communist coastal supply route. (US Navy)

reached the communist forces in the south to enable them to prosecute the war is self-evident; how much was lost along the way and what might have been the consequences had it arrived are questions likely to remain unanswered. The probability is that Giap had formed an accurate early estimate of the losses which would be incurred in transit and adjusted the flow to allow for this.

Coastal shipping was also widely employed by the communists to transport supplies from North Vietnam to the Mekong Delta or the Cambodian coast, where the so-called Sihanouk Trail led across the Plain of Reeds into South Vietnam. The South Vietnamese Navy had inherited several coastal patrol vessels from the French and supplemented these with 190 armed and motorised junks which were used to police the inshore traffic lanes. When the Americans arrived in 1965 they set up their own coastal surveillance force, Task Force 115, codenamed Market Time, which was equipped with former US Coast Guard cutters and 25-knot Swift boats, the latter armed with four .50-calibre machine guns and grenade launchers. A river patrol force, Task Force 116, codenamed Game Warden, was also set up to patrol the main waterways of the Mekong Delta with specially designed craft with fibreglass hulls and water-propulsion drive. The principal activities of Game Warden craft were surveillance, intelligence gathering and close co-operation with units of the US Navy's special forces, the SEALS. Later, hovercraft were used in the Plain of Reeds, where their ability to operate over water or swampland was extremely useful, although they were noisy and imposed a heavy maintenance workload.

Thus far, few attempts had been made to employ inshore squadrons in offensive operations against the enemy in the Mekong Delta. This region, which contained 60 per cent of South Vietnam's population, had become a hotbed of communist activity and included large areas into which the ARVN no longer ventured, including the Rung Sat Special Zone, the Coconut Grove in Go Cong province, the Cam Son Secret Zone, the U Minh Forest on the west coast and the Seven Mountains region on the Cambodian border. It has been estimated that in 1966 there were 80,000 Viet Cong present in the

Left *Watchful infantry cross a waterway by means of an assault boat.* (USAMHI)

Below right *A Kaman HH-43F rescue and recovery helicopter hovers over a PBR during Game Warden operations.* (US Navy)

Below *A river patrol boat (PBR) makes a high-speed run on a Delta river near Binh Thuy, June 1968.* (US Navy)

Mekong Delta, including 20,000 regulars, against whom the Saigon government had ineffectually deployed some 40,000 troops. In June 1967, however, the Americans established the Mobile Riverine Force by combining the Navy's River Flotilla One with the Army's 2nd Brigade 9th Infantry Division. The Riverine Force was equipped with armoured Assault Support Patrol Boats (ASPBs) armed with a 20mm cannon, a .50-calibre machine gun and a mortar, and a series of vessels converted from Landing Craft Mechanized. This included monitors for fire support, armed with a 40mm cannon or 105mm howitzer, a 20mm cannon, two .50-calibre and two .30-calibre machine guns, and an 81mm mortar; armoured troop carriers with one 20mm cannon and two .50-calibre machine guns; and command craft. Artillery support was provided by 105mm howitzers mounted on barges which were towed to their firing position; an alternative was to lower airmobile prefabricated firing

platforms into shallow water and jack them level. Most of the Riverine Force lived aboard converted tank landing ships which served as barracks.

The function of the Riverine Force was, in conjunction with other forces, to hunt down and destroy Viet Cong units amid the complex network of waterways which constituted the Mekong Delta. While the objective was hit by artillery and air strikes, the assault squadron would approach in column, preceded by ASPBs which served as minesweepers. Artillery and air support would cease as the column neared the landing area, which would be taken under direct fire by the ASPBs and monitors. The troop carriers would then turn together into the shore and land their assault infantry. As the latter fought their way inland, airmobile infantry would be lifted into blocking positions behind the enemy. Naturally, the communists attempted to disrupt such operations by mining channels and mount-

Above *US Navy monitor patrols the Quan Gong Tyon river in the Mekong Delta, April 1968. The vessel's armament consists of 40mm and 20mm cannon and .50-calibre machine guns.* (US Army)

Below *Although the infantrymen aboard this naval monitor seem relaxed, the danger of ambush was always present and the gun turrets have been traversed to port and starboard.* (USAMHI)

Above *M102 105mm howitzer of III/34th Artillery firing from a moored barge in support of 9th Division operations in the Mekong Delta.* (USAMHI)

Right *Northrop F-5 Freedom Fighter makes a precision bombing attack on a Viet Cong controlled area in the Mekong Delta.* (USAF).

ing ambushes from the banks, but by the end of 1968 their grip on the region had been so loosened that Route 4, closed for four years, could be opened again. The final phase of American naval participation in the river war was codenamed Sealords and involved the use of Market Time, Game Warden and Riverine Force units as appropriate to interdict waterways leading from Cambodia into South Vietnam.

Off the coast, American aircraft carriers launched countless air-support missions while, closer inshore, other warships provided a formidable weight of naval gunfire support for ground operations. This reached a climax in October and November 1968 when the battleship *New Jersey* fired over 3,000 rounds of 16-inch and 7,000 rounds of 5-inch ammunition against targets in the Demilitarized Zone.

Nevertheless, despite the application of so much science and technology, highly effective preventive medicine, efficient casualty evacuation and expert treatment of wounds, the jungle remained a place in which the soldier still relied on his primitive instincts for survival. Troops in the vicinity of Viet Cong base would find its approaches laced with booby traps which included pressure release mines, staked pits and trip wires which activated grenades, bows and heavy spiked balls concealed among the foliage. In these circumstances the presence of Viet Cong defectors, known as Kit Carson Scouts, was extremely welcome, since they knew how the enemy's mind worked and where he was likely to site such devices. Even when a base was penetrated there was little or nothing to see and the first warning would be a burst of fire from a camouflaged weapon slit.

Many Viet Cong and NVA sanctuaries

Monitors and armoured troop carriers (ATCs) of Naval Task Force 117 leave their base on the My Tho River for an operation upstream. (US Army)

existed in underground tunnel complexes, the greatest number of which were located between Saigon and the Cambodian border, particularly around Cu Chi and Ben Cat. Here the subsoil consists of hard laterite clay which retains its stability regardless of heavy rains, does not crumble when dug, and is therefore ideal material through which to tunnel. Some of the complexes had been started in the days of the French, but the firepower available to the Americans had itself ensured that they were expanded and many fresh complexes dug. Much careful thought and infinite labour had gone into their construction. They incorporated brief-ing rooms, sleeping quarters, latrines, storage chambers, hospitals, kitchens with remote smoke outlets, lookout posts and fire positions, ventilation shafts and wells dug down to the water table. These were con-nected by crawl tunnels on several levels which were separated by blast, gas and water proof trap doors. There would be several concealed entrances on the surface or below water level in a river bank. Often false tun-nels would be dug to confuse intruders, sometimes containing a booby trap or a poisonous snake, tethered to the wall, or a trip wire which opened a box of scorpions; natural inhabitants which discouraged visi-tors of their own accord included rats, bats, fire ants, centipedes and spiders. Life in the tunnels, with their stink and confinement, lack of air and hot darkness penetrated only by wick lamps, was an ordeal for the fittest and most balanced of men, but for the wounded and claustrophobic it must have been a living nightmare.

The tunnels presented the Americans with unexpected problems and special com-panies known as Tunnel Rats were formed to deal with them. Because of the psychological aspects of the work, only volunteers were required to perform it. Once an entrance had been discovered, smoke grenades would be dropped down the shaft and some idea of the extent of the complex could be gauged as the smoke began filtering out of concealed exits elsewhere. The exits would then be sealed and acetylene gas pumped in. This was detonated by explosive charge but experience revealed that the trap doors con-tained the blast, the damage being confined to the upper level. Better results were obtained by Tunnel Rats penetrating the complex, armed with a torch and pistol, trailing a rope with which they could be hauled to safety if they ran into trouble; despite every precaution, some were killed with bamboo spears as they changed levels through the trap doors, but they took a heavy toll in return. The only certain method by which a complex could be des-troyed was a B-52 strike, the hail of bombs creating an earthquake effect which crushed the tunnels and subterranean chambers. However, of the 500 miles of tunnels which existed, comparatively few were eliminated in this way.

The war continued to be waged at the tactical level until January 1967, when General Westmoreland mounted a corps-sized operation codenamed Cedar Falls, aimed at the destruction of the communist forces in an area known as the Iron Triangle. This lay only 18 miles to the north of Saigon and was formed by the junction of the Thi Tinh and Saigon Rivers, the third side fol-lowing a line across country from Ben Cat in the east to Ben Suc in the west. The whole area was covered in dense jungle in which the guerillas had located base camps, depots and headquarters in tunnel warrens, and since they were indistinguishable from the rest of the population their units were able to penetrate Saigon without difficulty. Of the Iron Triangle Westmoreland wrote,

The Vietnamese, before we arrived, would never dare go in there because it was totally dominated by the enemy. You couldn't go in with companies or battalions; they would have been chewed up, ambushed and deci-

193

mated. It took a massive troop effort to go there with safety and get the job done with minimum losses.

Cedar Falls contained three phases. First, the Iron Triangle would be sealed off; then it would be split in half; finally, each portion would be thoroughly searched. The operation began on 8 January when the eastern and western flanks were secured respectively by the US 1st and 25th Infantry Divisions, the former with elements of 173rd Airborne Brigade and 11th Armored Cavalry under command. Simultaneously, the remainder of 11th Armored Cavalry seized Ben Cat while I/26th Infantry launched a successful airmobile assault on Ben Suc. Next day the 11th Armored Cavalry, less one squadron, drove across the Triangle, cutting it in two. The search and destroy teams moved in, assisted by tankdozers where necessary. Many of the guerrillas managed to evade their cordons, but others were killed while attempting to escape. Most clashes involved platoon-sized groups of Viet Cong but, overall, fighting was light. When the operation ended on 25 January some 750 guerrillas had been killed and 280 captured. Included in the material captured were 23 heavy weapons, 590 small arms, large quantities of ammunition and uniforms, enough rice to feed a division for a year and, most important of all, 500,000 pages of documents which revealed the entire Viet Cong and NVA order of battle. The Allied loss was 72 Americans killed and 337 wounded, 11 ARVN personnel killed and eight wounded, one tank and three APCs destroyed and three tanks, nine APCs and two helicopters damaged.

When the Viet Cong began filtering back into the Iron Triangle in early February they found the region transformed. The villages on which they relied for support had been destroyed and their inhabitants evacuated, thereby creating a Free Fire Zone for artillery and air strikes. Much of the con-

cealment offered by the jungle had also gone as a result of defoliation and the carving of lanes by Rome ploughs.

Cedar Falls was followed by Operation Junction City, which lasted from 22 February until 14 May and was aimed generally at the destruction of another communist sanctuary, War Zone C, occupying a 30- by 50-mile area of Tay Ninh province on the Cambodian border, and specifically at the elimination of the communist Central Office of South Vietnam (COSVN), the headquarters through which Hanoi conducted the war in the south. The troops committed were those previously involved in Cedar Falls, reinforced by brigades from the US 4th and 9th Infantry Divisions and ARVN Ranger and Marine battalions. Heavier resistance was encountered but COSVN was forced to withdraw into Cambodia and its activities were seriously disrupted for a while. Over 2,700 guerrillas were killed and 200 were captured, arrested or defected. Some 500 weapons were also captured, together with ammunition, 850 tons of rations and another 500,000 pages of documents. American casualties amounted to 282 killed and 1,576 wounded, plus 3 tanks, 5 howitzers and 21 APCs destroyed. During the operation the 173rd Airborne Brigade's II/503rd Infantry made the only major parachute insertion of the war, dropping into a jungle clearing four miles from the Cambodian border.

The Americans were now imposing a rate of attrition which would have been regarded as unacceptable by any other than a communist army, although the enemy always seemed able to fill the gaps in his ranks and his ability to merge into the villages, sink into the ground or vanish across the border was frustrating in the extreme. Even more insidious was the effect on morale produced by the strident anti-war movement in the United States and the West. This was an era when left-wing liberal attitudes,

An F-100 Super Sabre provides close support during Operation Junction City. (USAF)

however illogical, had become fashionable, particularly in universities and in the expanding television industry, which had an immense influence on public opinion. In such circles the involvement of the United States and her allies in what was seen as a purely Vietnamese problem was regarded as being immoral, despite the fact that it was the NVA, with Soviet and Chinese backing, which had invaded South Vietnam. Again, while the war correspondents of the Second World War and Korea were experienced professionals, many of those reporting events in Vietnam were callow and prejudiced. The public was therefore exposed to nightly television programmes which showed the full horror of the war without revealing the operational background. Thus, while footage of Cedar Falls showed villages put to the torch and their inhabitants transported unwillingly to camps, neither the underlying reasons nor the fact that it was the communists who chose to fight among the civilian population were

satisfactorily explained. It is now accepted that the ultimate communist victory in South Vietnam was won on the television screens of the United States and the Free World.

There was, however, another side to the coin. In Hanoi the agitation caused by the peace movement was viewed with justifiable satisfaction, and it seemed that world opinion was moving against the Americans just as it had against the French prior to Dien Bien Phu. Furthermore, exaggerated claims of success from NVA commanders in the south, no doubt prompted by considerations of face, suggested that the anti-communist forces were being roundly beaten at every turn. It was reported, for example, that during Junction City the Allied losses included 13,500 killed and 1,000 vehicles. There were, too, wildly optimistic suggestions that the population was ready to rise *en masse*, given the chance. The communist leadership therefore became victims of their own propoganda and, although Giap and the

195

Chinese thought it premature, the decision was taken to activate the third phase of revolutionary warfare, the capture of the towns and cities of South Vietnam.

After careful preparation the offensive began on 30 January 1968, timed to coincide with the sacred festival of Tet, the lunar new year, during which both sides had previously observed a truce. In 5 major cities and 36 provincial capitals the Viet Cong and NVA attempted to seize control. To their horror, there was no popular rising in their support and they suddenly found themselves exposed to street fighting in which the Allies were able to deploy their immense firepower. Estimates of communist dead vary between 30,000 and 50,000 but the true figure will probably never be known. Allied casualties amounted to 4,000 killed, 16,000 wounded and 600 missing. Some 38,000 civilians had been killed or wounded in the fighting, many political opponents being murdered by the Viet Cong during the early stages, and 630,000 were left homeless. The majority of attacks were defeated within 24 hours, although at Hue fighting continued around the embattled citadel until 24 February. In

the military sense, the Tet offensive was an unmitigated disaster for the communists and resulted in the virtual destruction of the Viet Cong—not that Hanoi was unduly saddened by this, since it had no intention of sharing power with the southerners. In the political sphere, however, naive reporting by shocked journalists handed the NVA a victory on a plate. Instead of treating Tet as a major communist set-back brought about by serious miscalculation, correspondents viewed it as a frightening demonstration of Hanoi's ability to respond at a high level despite all that had gone before. The unease generated by this distorted perspective caused many Americans to question whether the war could be won, and if it could not, should not the United States withdraw from the conflict at the earliest possible opportunity? Thus, while Tet was a resounding military defeat, it marked a change in attitudes from which Hanoi was to benefit in the long term.

One aspect of the fighting which continued to absorb the public's attention was the 77-day siege of the isolated US Marine base at Khe Sanh, 14 miles south of the Demilitarized Zone, which had begun on 21

Left *Look and listen. A 5th Infantry Division patrol pauses to investigate a suspicious contact.* (USAMHI)

Right *The M60 General Purpose Machine Gun incorporated some features of the famous German MG42 and could be fired either from a bipod, or, as here, from a specially designed tripod.* (USMC)

January. Giap later denied that he intended turning the base into another Dien Bien Phu, but if the objective was of no importance to him he would hardly have detailed the NVA's regular 304th and 325th Divisions to capture it in the largest set-piece battle since the days of the French. The probability is that its fall was intended to crown the other achievements of Tet, so demonstrating that the Americans could be defeated in the field.

Khe Sanh was situated on a plateau below the Dang Tri Mountains, overlooking a tributary of the Quang Tri River. Like Dien Bien Phu, it incorporated a runway and was surrounded by hills on which outposts were established to prevent the enemy dominating the interior; two of these, Hills 881 South and 861, incorporated fire bases. The garrison, commanded by Colonel David Lownds, consisted of the 26th Marine Regiment, reinforced with I/9th Marines and the ARVN 37th Ranger Battalion. The defences contained three batteries of 105mm howitzers and one battery of 155mm howitzers, plus 90mm anti-tank guns, all manned by Marines, and had the additional support of

four batteries of Army 175mm guns, three firing from Camp Carroll to the east and one from a base known as The Rockpile to the north. Also present were six M48 tanks, ten Ontos anti-tank vehicles, each mounting six 106mm recoilless rifles, and four Duster self-propelled automatic anti-aircraft guns which could be used in the ground role. Some miles to the west on Route 9 lay the Lang Vei Special Forces Camp, held by 24 Green Berets and 900 *Montagnard* irregulars.

Shortly after midnight on 21 January an attack on Hill 861 was repulsed, but at first light rocket, mortar and artillery fire damaged the runway and hit the main ammunition dump, sending 1,500 tons of explosives skywards in a thunderous series of explosions. The NVA continued shelling the base and sapping towards it, but on 5 February major assaults on Hills 881 South and 861 were thrown back. During the night of 7 February, however, a communist attack supported by a dozen Soviet-built PT-76 light tanks penetrated the perimeter of Lang Vei Special Forces Camp. Several tanks were knocked out before the camp was overrun the following morning; 14 of the Green

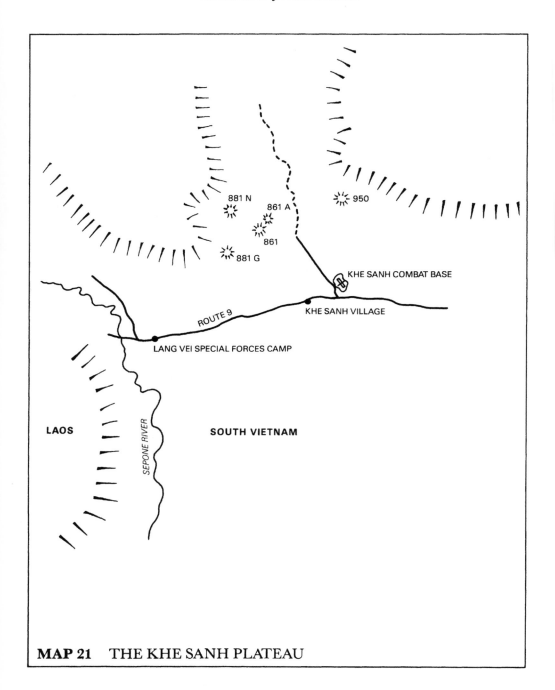

MAP 21   THE KHE SANH PLATEAU

Berets and several hundred *Montagnards* managed to reach Khe Sanh through an intervening jungle. On 9 February I/9th Marines defeated an attack on one of their outposts west of the perimeter and after this the pressure eased although the shelling continued. This reached its climax on 23 February, when over 1,300 shells hit the base, destroying another ammunition dump. An attempt to break into the sector held by the ARVN Rangers early on 1 March foundered in a hail of exploding Claymores and automatic fire which left 70 bodies tangled in the wire. It was anticipated that the enemy would attempt to mark the anniversary of Dien Bien Phu on 13 March with a mass assault, but his concentrations were broken up by artillery fire and a B-52 strike. A similar attack seemed likely during the night of 21/22 March, when a third ammunition dump was destroyed, critically reducing the number of shells available; the threat

diminished when an AC-47 gunship began circling methodically overhead. By the end of the month the NVA had withdrawn its 325th Division and clearly abandoned any hope of capturing the base. On 6 April Khe Sanh was relieved by Operation Pegasus, involving the 1st Cavalry Division and 1st Marine Regiment in a joint airmobile/ground assault along Route 9 from the east.

Khe Sanh had never been in any real danger. In contrast to the NVA's average daily ammunition expenditure of 150 rounds, the Americans fired almost 159,000 rounds throughout the siege, supplemented by 96,000 tons of bombs. Although several transport aircraft and helicopters were lost, supplies continued to flow into the base, accurately para-dropped whenever the runway was out of action, the outposts being maintained by means of a helicopter shuttle service. American casualties amounted to 199 killed and 1,600 wounded. The lowest

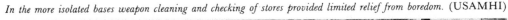

In the more isolated bases weapon cleaning and checking of stores provided limited relief from boredom. (USAMHI)

estimate of the NVA's losses puts these at 9,800, of whom one-third were killed. The most notable aspect of the defence was the use of isolating box barrages which took advantage of the NVA's tendency to attack with their battalions in column and isolated the assault elements from their reserves. Three of the garrison's four batteries formed a three-sided box around the leading enemy battalion. The fourth battery then dropped a rolling barrage on the open side, i.e. that closest to the defences, and this was walked up and down the box. If any of the attackers emerged, they immediately came under fire from the perimeter defences. Simultaneously, the four Army 175mm batteries formed the two sides of an outer box, 500 yards outside the inner box, and walked their shells across the intervening area; the third side of the outer box, which faced the enemy, was closed by tactical air strikes.

In June 1968 General Creighton W. Abrams took over from Westmoreland, who had been appointed Army Chief of Staff. The war continued much as before, although its course would be influenced more by political considerations than battlefield decisions. The Americans had indeed solved many of the problems associated with jungle warfare and they were inflicting heavy casualties, but pressures on the government demanded disengagement and a progressive handing over of the conduct of the war to the ARVN. Unfortunately, the ARVN as a whole continued to perform badly, was unpopular with its own people, and still relied heavily on American support and advice.

Secret talks between the Americans and the North Vietnamese commenced in Paris on 13 May 1968. The aims of the former included a ceasefire, an honourable disengagement and the release of prisoners held by Hanoi; those of the latter an end to the bombing of North Vietnam and the withdrawal of the United States and her allies. Communist intransigence ensured that the

negotiations lasted years rather than months, and during this time the Allies steadily reduced the number of troops serving in Vietnam to a fraction of their former levels. Ho Chi Minh died on 4 September 1969, but this in no way diminished the NVA's level of activity in the south. In fact, taking advantage of the reduced American presence, it launched a conventional offensive into South Vietnam in January 1972. This was ineptly conducted and was defeated by a combination of American air power and a much improved ARVN. At this point China again intervened. For a while it had suited her that so much of America's strength had been tied down in South Vietnam, but as a result of worsening relations with the Soviet Union she was now anxious to be on better terms with the West. The threatened withdrawal of Chinese support, coupled with intensified American bombing, led directly to a ceasefire being concluded on 23 January 1973. On March 29 the last American troops left Vietnam and on 1 April Hanoi released the last of its American prisoners.

The peace was short-lived. In January 1974 President Thieu of South Vietnam announced that communist troops were again active throughout the country. Later that year the United States Congress placed severe limits on the amount of military aid which would be given to South Vietnam and effectively prohibited further American intervention in Indochina by the Case-Church Enactment, which was endorsed by the House of Representatives. It was the signal Giap had been waiting for, the signal that proved conclusively that the American political will had been broken. On 5 March 1975 the NVA launched a full-scale invasion of South Vietnam. Significantly, it employed not the tactics of revolutionary warfare, but a conventional blitzkrieg advance spearheaded by armour, and had clearly learned much from its mistakes in 1972. The South

—Field Artillery 1954—1973, Department of the Army, Washington

Padden, Ian, *U.S. Rangers*, Bantam

Perrett, Bryan, *Tank Tracks to Rangoon*, Robert Hale

A History of Blitzkrieg, Robert Hale

Phillips, C.E. Lucas, *The Raiders of Arakan*, Heinemann

Purnell's *History of the First World War*

History of the Second World War

Rooney, David, *Stilwell*, Pan/Ballantine

Rutherford, Ward, *Fall of the Philippines*, Pan/Ballantine

Ste Croix, Philip de, *Airborne Operations*, Salamander

Shores, Christopher, *Ground Attack Aircraft of World War II*, Macdonald and Jane's

Simpson, Charles M., *Inside the Green Berets — A History of the US Army Special Forces*, Arms & Armour Press

Smith, Peter C., *Jungle Dive Bombers At War*, John Murray

Starry, General Donn A., *Armoured Combat in Vietnam*, Blandford

Stewart, Adrian, *The Underrated Enemy— Britain's War With Japan December 1941- May 1942*, William Kimber

Swinson, Arthur, *Mountbatten*, Pan/Ballantine

Thompson, Sir Robert, and others, *War In Peace—An Analysis of Warfare Since 1945*, Orbis

Tolson, Lieutenant-General John J., *Vietnam Studies—Air Mobility 1961-1971*, Department of the Army, Washington

Vader, John, *New Guinea—The Tide Is Stemmed*, Pan/Ballantine

Warner, Philip, *The SAS*, William Kimber

Windrow, Martin and Braby, Wayne, *French Foreign Legion Paratroops*, Osprey

Wolff, Leon, *Little Brown Brother—America's Forgotten Bid For Empire*, Longmans

INDEX

(Figures in *italics* refer to illustrations)

204